G. Csomós   J. Fehér (Eds.)

Free Radicals and the Liver

With Contributions by
E. Albano, B.R. Bacon, F. Biasi, J. Blanck, A. Blázovics, W. Bors,
R.S. Britton, E. Chiarpotto, G. Csomós, O. Danni, M.U. Dianzani,
E. Fehér, J. Fehér, E.A. Glende Jr., J. Györgi, W. Heller,
V.E. Kagan, H. Kappus, C. Michel, R. O'Neill, L. Packer, G. Poli,
R.O. Recknagel†, H. Rein, O. Ristau, K. Ruckpaul, M. Saran,
E.A. Serbinova, H. Toncser, A. Vereckei

With 85 Figures

Springer-Verlag
Berlin Heidelberg New York
London Paris Tokyo
Hong Kong Barcelona
Budapest

GÉZA CSOMÓS, M.D.
Liver Research Foundation
Semmelweis University of Medicine
Szentkirályi u. 46
1088 Budapest
Hungary

JÁNOS FEHÉR, M.D.
Second Department of Internal Medicine
Semmelweis University of Medicine
Szentkirályi u. 46
1088 Budapest
Hungary

ISBN 3-540-54445-3 Springer-Verlag Berlin Heidelberg New York
ISBN 0-387-54445-3 Springer-Verlag New York Berlin Heidelberg

Library of Congress Cataloging-in-Publication Data. Free radicals and the liver / G. Csomós,
J. Fehér (eds.); with contributions by E. Albano . . . [et al.]. p. cm. Includes bibliographical
references.
ISBN 3-540-54445-3. — ISBN 0-387-54445-3
1. Liver—Pathophysiology. 2. Free radicals (Chemistry) 3. Liver cells. 4. Membrane lipids
—Peroxidation. I. Csomós, G. (Géza) II. Fehér, J. (János), 1932– . III. Albano, E.
[DNLM: 1. Free Radicals—toxicity. 2. Liver Diseases—chemically induced. WI 700 F8528]
RC846.9.F74 1992 616.3′62071—dc20 DNLM/DLC for Library of Congress 92-2150 CIP

© Springer-Verlag Berlin Heidelberg 1992
Printed in Germany

Typesetting: Best-set Typesetter Ltd., Hong Kong

21/3130-5 4 3 2 1 0 – Printed on acid-free paper

Preface

In recent years numerous research teams studying free radicals in the field of medicine and have shown, using various methods, that free radicals play an important role in toxic liver disorders. For this reason we chose this as the main theme for the President's Meeting of the EASL held in Budapest.

This book aims to present a balanced, clinical portrayal of current scientific knowledge. It draws on accounts from observations, experiences, and evaluations from various fields, since many questions regarding the pathogenetic mechanism of the liver have been addressed in recent years, both clinically and experimentally, by our working group.

For the authors, apart from pathophysiological changes by free radicals, a particular matter of concern was also to highlight the therapeutic possibilities.

The report supports these statements with new scientific findings and gives practical advice for both research and clinical physicians.

G. Csomós
J. Fehér

Contents

List of Contributors

ALBANO, E.
Department of Experimental Medicine and Oncology
of the University
Corso Raffaello 30
10125, Torino, Italy

BACON, B.R.
Division of Gastroenterology and Hepatology
St. Louis University School of Medicine
3635 Vista Ave. at Grand Blvd.
St. Louis, MS 63110-0250, USA

BIASI, F.
Department of Experimental Medicine and Oncology
of the University
Corso Raffaello 30
10125, Torino, Italy

BLANCK, J.
Central Institute of Molecular Biology
Academy of Sciences
O-1115 Berlin-Buch, FRG

BLÁZOVICS, A.
Second Department of Medicine
Semmelweis University Medical School
Szentkirályi u. 46
1088 Budapest, Hungary

BORS, W.
Institut für Strahlenbiologie
GSF Research Center
W-8042 Neuherberg, FRG

BRITTON, R.S.
Department of Medicine
Section of Gastroenterology
Louisiana State University Medical Center
Shreveport, LA 71130, USA

CHIARPOTTO, E.
Department of Experimental Medicine and Oncology
of the University
Corso Raffaello 30
10125, Torino, Italy

CSOMÓS, G.
Liver Research Foundation
Semmelweis University of Medicine
Szentkirályi u. 46
1088 Budapest, Hungary

DANNI, O.
Institute of General Pathology of the University
Via Padre Manzella 4
07100, Sassari, Italy

DIANZANI, M.U.
Department of Experimental Medicine and Oncology
of the University
Corso Raffaello 30
10125, Torino, Italy

FEHÉR, E.
First Department of Anatomy
Semmelweis University Medical School
Szentkirályi u. 46
1088 Budapest, Hungary

FEHÉR, J.
Second Department of Internal Medicine
Semmelweis University of Medicine
Szentkirályi u. 46
1088 Budapest, Hungary

GLENDE, E.A. Jr.
Department of Physiology and Biophysics
School of Medicine
Case Western Reserve University
Cleveland, OH 44106, USA

GYÖRGI, J.
Second Department of Internal Medicine
Semmelweis University of Medicine
Szentkirályi u. 46
1088 Budapest, Hungary

HELLER, W.
Institut für Biochemische Pflanzenpathologie
GSF Research Center
W-8042 Neuherberg, FRG

KAGAN, V.E.
Department of Molecular and Cell Biology
251 LSA
University of California
Berkeley, CA 94720, USA

KAPPUS, H.
Department of Dermatology
Rudolf Virchow Clinic
Free University of Berlin
Augustenburger Platz 1
W-1000 Berlin 65, FRG

MICHEL, C.
Institut für Strahlenbiologie
GSF Research Center
W-8042 Neuherberg, FRG

O'NEILL, R.
Division of Gastroenterology and Hepatology
St. Louis University School of Medicine
3635 Vista Ave. at Grand Blvd.
St. Louis, MS 63110-0250, USA

PACKER, L.
Department of Molecular and Cell Biology
251 LSA
University of California
Berkeley, CA 94720, USA

POLI, G.
Department of Experimental Medicine and Oncology
of the University
Corso Raffaello 30
10125, Torino, Italy

RECKNAGEL, R.O.†
Department of Physiology and Biophysics
School of Medicine
Case Western Reserve University
Cleveland, OH 44106
USA

REIN, H.
Central Institute of Molecular Biology
Academy of Sciences
O-1115 Berlin-Buch, FRG

RISTAU, O.
Central Institute of Molecular Biology
Academy of Sciences
O-1115 Berlin-Buch, FRG

RUCKPAUL, K.
Central Institute of Molecular Biology
Academy of Sciences
O-1115 Berlin-Buch, FRG

SARAN, M.
Institut für Strahlenbiologie
GSF Research Center
W-8042 Neuherberg, FRG

SERBINOVA, E.A.
Department of Molecular and Cell Biology
251 LSA
University of California
Berkeley, CA 94720, USA

TONCSER, H.
Second Department of Internal Medicine
Semmelweis University of Medicine
Szentkirályi u. 46
1088 Budapest, Hungary

VERECKEI, A.
Second Department of Internal Medicine
Semmelweis University of Medicine
Szentkirályi u. 46
1088 Budapest, Hungary

Role of Free Radical Reactions in Liver Diseases

J. Fehér, G. Csomós, and A. Vereckei

Introduction

Free radical reactions play a significant role in toxic liver injuries. Two traditional groups of liver injury induced by drugs and chemicals can be distinguished: (1) the direct toxic type and (2) the idiosyncratic type. Liver injury of the direct toxic type generally develops following toxin exposure; it is dose dependent, the incubation period is short, and the injury frequently affects other organs (e.g., kidney). Direct toxins frequently cause typical zonal necrosis usually without concomitant signs of hypersensitivity. The typical idiosyncratic reaction appears only during a shorter period of exposure, it cannot be predicted, it is not dose dependent, its incubation period varies, it is sometimes (in one-fourth of cases) accompanied by extrahepatic symptoms of hypersensitivity (fever, leukocytosis, eosinophilia, rashes), and its morphological picture shows great variety. A proportion of direct toxins are toxic themselves; in the other proportion the basic compound is not toxic but changes into toxic metabolites in the liver.

Immune reactions have been presumed to play a role in the pathogenesis of idiosyncratic reactions, this being indicated by the associated hypersensitivity symptoms. However, it has been shown that the majority of reactions of the idiosyncratic type correspond to direct hepatotoxicity, which is caused mainly by toxic metabolites of the inducing agent and not by the basic compound itself. A small proportion of idiosyncratic reactions may really be of immunological origin, when the drug, such as haptene, on connection with some protein induces an immune response; these reactions occur only in the case of repeated administration of the drug. It is not exactly known why certain toxic metabolites cause liver injury only in some cases, but the existence of chemical environment-dependent, alternative metabolic pathways (e.g., for halothane, an oxidative or reductive metabolic route exists depending on tissue oxygen tension), metabolizing depending on the drug acetylator state (e.g., iproniazid) and individual differences of efficiency of the protective system (against toxic metabolites, antioxidants, cytochrome system) might play some role in the induction of such damage.

Recently, it has been found that most toxic metabolites originating from drugs and chemicals are very reactive free radicals, having an unpaired electron in their external electron shell (corresponding to free valence) which can damage tissues and cells reacting with the surrounding macro-molecules and membranes, forming new free radicals as a chain reaction [1–4].

The liver is well protected against free radicals developing in the organism; it is one of our best antioxidant-supplied organs, which is probably due to one of the major functions of liver, namely detoxication of drugs, chemicals, and toxic materials, with subsequent release of free radicals. This is proved by the fact that in normal bile peroxidized lipids which were produced by free radical chain reactions can also be detected [5, 6].

It should be emphasized that free radical reactions are not solely of a damaging character but occur in the organism in a controlled way and can be the basis of several physiological processes, e.g., phagocytosis, arachidonic acid metabolism, regulation of certain immune processes [7, 8].

The pathological free radical reactions and one of their sequelae, lipid peroxidation (LPO), do not necessarily cause cell and tissue damage. Antioxidant protection of cells and tissues is able to prevent free radical injury and enables the already developed damage to be reversed. According to recent investigations, LPO caused by free radical reactions, or covalent binding of radical products to biomolecules does not lead directly to cellular destruction, but only indirectly, via further reactions. Such intermediate steps can be phospholipase A_2 activation, accumulation of lysophosphatides, and poly-ADP-ribose polymerase repair enzyme activation, following oxidative damage of DNA, with subsequent NAD and ATP depletion. This may mean that the irreversible cellular and tissue damage can be prevented perhaps not only by administration of antioxidants, but also by compounds (e.g., phospholipase A_2 inhibitors) affecting the above-mentioned biochemical mechanisms [9–12].

Alcoholic Liver Damage

In order of gravity, three forms of alcoholic liver damage can be distinguished: (1) steatosis, (2) alcoholic hepatitis, and (3) alcoholic cirrhosis. Probably, LPO plays a casual role in all three forms; however, experimental evidence proves it to be present mainly in steatosis. A decreased hepatic GSH (glutathione, reduced form) content and an increased conjugate diene content were found in liver biopsy samples of patients with alcoholic liver disease [13, 14]. Following ethanol inhalation in rats, decreased hepatic catalase (CAT) and CuZn-SOD (superoxide dismutase) activity were found with normal Mn-SOD activity. The observed differences are probably explained by reactive oxygen intermediates (ROIs) released during microsomal metabolism of ethanol [15]. Also in rats and baboons, both acute and chronic alcohol administration caused an increase in conjugate diene and a significant decrease in GSH level in the liver [16]. The LPO mechanism

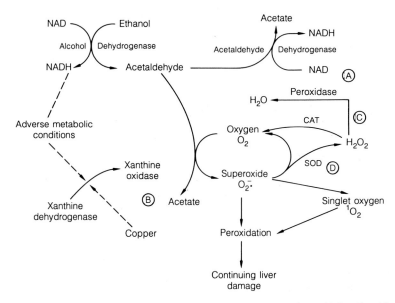

Fig. 1. Hypothetic free radical pathogenesis of alcoholic liver injury. (After Lewis)

induced by alcohol is not exactly known. It can possibly be explained by Lewis and Paton's hypothesis, which presumes cytoplasmatic xantine oxidase to play a role (Fig. 1).

The first metabolite of alcohol, acetic aldehyde, can be metabolized not only by acetic aldehyde dehydrogenase (though this enzyme metabolizes it mainly) but also by xantine oxidase.

In healthy cells xantine oxidase acts and NAD^+-reducing dehydrogenase, not as oxidase. As a result of unfavorable metabolic circumstances, e.g., proteolysis, ischemia, however, it changes into oxidase. The enzyme may use acetic aldehyde instead of hypoxantine as substrate and, while it oxidizes it to acetate, $O_2^{\cdot-}$ is produced from molecular oxygen. As acetic aldehyde is present in a large scale in chronic alcoholics, excessive $O_2^{\cdot-}$ production develops because an acetic aldehyde supply (in contrast to purine metabolites) is unlimited.

The hypothesis of Lewis and Paton is supported by experimental results; if tolbutamide, which inhibits aldehyde dehydrogenase, was given to rats, it accelerated the decrease in GSH levels induced by alcohol; it therefore increased the levels of circulating acetic aldehyde, while allopurinol had a reverse effect. In vitro bovine xantine dehydrogenase oxidase incubated with acetic aldehyde was somewhat different from the effect expected on the basis of the Lewis-Paton theory. At the same time the acetic aldehyde did not influence the activity of enzyme oxidase. Consequently, the ROIs produced by acetic aldehyde do not originate from the metabolization of acetic aldehyde by xantine oxidase, but they are the sequelae of dehydrogenase oxidase conversion [17, 18].

Another possible hypothesis is that the consumption of ethanol increases the activity of microsomal enzymes of the liver, also including the system of oxidizing microsomal ethanol, which participates in the microsomal metabolism of ethanol. ROIs released during microsomal enzyme activity may have a role in the effect of alcohol-producing hepatic LPO [15, 16, 19]. Besides, acetic aldehyde produces ROIs, not only in course of the xantine oxidase dependent metabolism presumed in the Lewis-Paton hypothesis, but also in that regulated by the aldehyde oxidase enzyme [15, 20].

The binding of acetic aldehyde to hepatic steatosis (HS)-containing GSH precursors results in a decrease of the GSH concentration and is a third alternative mechanism [21]. Increased CuZn-SOD activity was found in the erythrocytes of chronic alcoholics suffering from liver damage. So, besides other known tests, this examination seems to be suitable for the detection of alcohol consumption [22].

The free radical reactions play a role in the pathogenesis of some hematological disorders associated with alcoholic liver diseases (macro-cytosis, hemolytic anemia, formation of target cell and spur cell). In patients with the Zieve syndrome (alcoholic liver disease associated with hyper-lipidemia and hemolytic anemia), a decreased number of erythrocytes, polyunsatured fatty acid, GSH and vitamin E content, decreased pyruvate kinase activity, and a lower serum vitamin E level were found. The erythrocytes of patients were more sensitive to H_2O_2 stress in vitro. Thus besides other known factors (folic acid deficiency, decreased prothrombin level), the disbalance of the oxidant-antioxidant system also plays a major role in the induction of complications of hematopoiesis in alcoholic liver disease [23].

Our own investigations also support the role of free radical reactions in the pathogenesis of chronic liver diseases. Lysosomal enzyme activity was studied in serum and granulocytes of patients with chronic liver disease. It is known that LPO can damage the lysosomal membranes and lysosomal enzyme release. In HS, chronic active hepatitis (CAH), and hepatic cirrhosis serum β-glucuronidase activity was significantly increased. Serum acid phosphatase activity was increased in all three diseases, but in CH this increase was not significant. Finally, the serum cathepsin D activity was not significantly decreased, while the activity of β-glucuronidase in the granulocytes of hepatopaths was depleted compared to the control group. Calculating the release rate of β-glucuronidase from the granulocytes, this parameter was higher in patients with chronic liver disease. Higher serum β-glucuronidase activity partly originates from the granulocytes, but it cannot be totally responsible for the greater activity of serum. It confirmed our suggestion that the enhanced enzyme release increases from different tissues, including liver, are a result of decreased stability of lysosomal membranes. Lysosome damage in CH was confirmed by electron microscopic examination. It is interesting that the level of serum β-glucuronidase was the highest in patients suffering from CH; a lower increase was found in CAH and SH. Accordingly, the β-glucuronidase activity of granulocytes definitely decreased in CH – while there was a smaller decrease in CAH and SH. Similarly to mitochondrial and microsomal liver enzyme examinations routinely applied in diagnostics, the serum and granulocyte lysosomal enzyme tests, respectively, can be used in diagnostics of liver diseases; moreover, their determination may have differential diagnostic significance in differentiation of chronic liver diseases [24–27]. Administration of the antioxidant cyanidanol-3 settled the elevated acid phosphatase level of hepatic patients with chronic disease within 3 months, significantly decreased β-glucuronidase activity, and normalized the enzyme release from granulocytes [28].

In our department the efficiency of silymarin (Legalon) treatment, which contains silymarin and has antioxidant properties besides other major pharmacological effects, was investigated in a double-blind study on patients with chronic alcoholic liver disease. Thirty-six patients with chronic alcoholic liver disease participated in this study (27 men, 9 women; average age, 46 ± 7 years). Daily alcohol consumption exceeded 60 g in men and 30 g in women. The period of chronic alcohol consumption was 8 ± 4 years. The patients were vascularly compensated, symptoms of encephalopathy were not observed, malnutrition did not occur, and there was no other associated disease. The virus and immunological (antinuclear antibody, anti-smooth muscle antibody) markers were negative. Silymarin-placebo randomization was performed by the pharmaceutical company Madaus, Cologne. The placebo contained the vehicle silymarin in the same form. The treatment lasted for 6 months.

The code was disclosed by the pharmaceutical company after the end of treatment. Seventeen patients (15 men, 2 women; average age, 48 ± 7 years) took 3 × 140 mg silymarin/day (3 × 1 Legalon tablets) and 19 patients (12 men, 7 women; average age, 44 ± 6 years) took 3 × 1 placebo tablets. Histological examination of the liver in the silymarin group verified micronodular cirrhosis, with signs of chronic inflammatory activity in six patients (associated with steatosis in four patients, hemosiderosis in three patients). In ten patients steatosis with reactive fibrosis and round-cell infiltration (associated with hemosiderosis in one patient) and in one patient septal fibrosis showing chronic inflammatory activity were observed. In the placebo group the histological diagnosis was micronodular cirrhosis with chronic inflammatory activity in 5 patients (associated with steatosis in 1 patient), steatosis with reactive fibrosis and round-cell infiltration in 12 patients (associated with hemosiderosis in 1 patient), acute alcoholic hepatitis associated with septal fibrosis in 1 patient, and septal fibrosis showing chronic inflammatory activity in 1 patient.

Table 1. Effect of silymarin treatment on lipid peroxidation and the antioxidant system in chronic alcoholic liver diseases, in a 6-month double-blind investigation (average ±SEM)

Groups (n)		MDA (nM/ml)	Serum GPX (U/g plasma protein)	Free SH (µM/ml)	Erythrocyte SOD (U/ml)	Lymphocyte SOD (U/ml)
Silymarin (17)						
I	0 months	15.1 ± 2.5	0.65 ± 0.28	0.44 ± 0.20	72.9 ± 14.5	32.6 ± 10.3
II	6 months	10.2 ± 1.0	0.94 ± 0.25	0.63 ± 0.17	130.8 ± 19.6	74.9 ± 19.3
Placebo (19)						
III	0 months	14.7 ± 2.3	0.67 ± 0.21	0.45 ± 0.12	76.5 ± 20.1	29.4 ± 14.2
IV	6 months	15.9 ± 2.1	0.54 ± 0.26	0.43 ± 0.15	85.7 ± 21.7	27.7 ± 16.1
Significance						
I vs. II		$p < 0.02$	$p < 0.05$	$p < 0.05$	$p < 0.001$	$p < 0.01$
II vs. IV		$p < 0.02$	$p < 0.02$	$p < 0.05$	$p < 0.01$	$p < 0.01$

Of the elements of the antioxidant protective system, serum glutathione peroxidase activity (GPX, EC.1.11.1.9) free SH groups in the serum, superoxide dismutase (SOD, EC.1.15.1.1) activity of erythrocytes and lymphocytes, and SOD expression in lymphocytes were determined.

Initial values and values following treatment of the two groups of patients (placebo and silymarin treatment, respectively) are summarized in Table 1. It was established that there were no major differences in the parameters examined (I vs. III) in the initial group data. The results showed no significant changes following the administration of placebo (III vs. IV).

The level of malondialdehyde (MDA) marker of serum LPO decreased significantly during the silymarin treatment ($p < 0.002$). There was a significant difference between the values of the two groups following the treatment ($p < 0.002$).

Following silymarin administration the serum GPX activity significantly increased ($p < 0.05$), and the values of groups following treatment showed a significant difference ($p < 0.02$). Silymarin treatment significantly increased the SH group level ($p < 0.05$). The values of the two groups following treatment also showed a significant difference ($p < 0.05$).

During silymarin treatment the SOD activity of erythrocytes and lymphocytes increased distinctly in both types of cells (erythrocyte SOD, $p < 0.001$; lymphocyte SOD, $p < 0.01$). The values of groups following the treatment were significantly different (erythrocyte SOD, $p < 0.01$; lymphocyte SOD, $p < 0.01$).

During 6 months administration of placebo, no marked changes in SOD expression of lymphocytes were observed, whereas the SOD expression of lymphocytes increased significantly in the group treated by silymarin (Fig. 2).

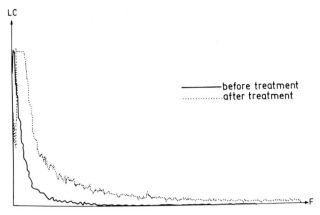

Fig. 2. Effect of 6-month silymarin (Legalon) treatment on lymphocyte SOD expression of chronic alcoholic hepatopaths. Fluorescence histogram. *LC*, lymphocyte count; F, intensity of fluorescence; ———, before treatment, · · · · · · · ·, following treatment

During the 6 months of treatment side effects ascribed to silymarin were not observed. Our studies showed that in patients with chronic alcoholic liver disease the extrahepatically detectable oxidative stress state – its typical biochemical parameters – were favorably influenced by Legalon treatment. Applied therapy decreased lipid peroxidation and improved the patients' antioxidant protection. Our studies also showed an immunomodulating effect of Legalon, though details are not given here. Following the 6-month treatment, all these favorable effects resulted in the settling or significant improvement of liver function values in patients, proving the liver-protective effect of the drug [29, 30].

Liver Injury Caused by Drugs and Chemicals

The best-known liver injury is caused by carbon tetrachloride (CCl_4) and by galactose amine. The prerequisite for their damaging effect is an actively metabolizing liver. CCl_4 is reduced by cytochrome P-450, a terminal oxidase of a heterogeneous oxidase system in the endoplasmic reticulum of liver. Following the splitting of the CCl_3-D1 bond, the compound is metabolized to $CCl_3{}^{\cdot-}$, then it forms a more electrophilic and reactive radical $CCl_3O_2{}^{\cdot-}$. The latter produces LPO of the endoplasmic reticulum, so in liver injuries caused by CCl_4 LPO develops mainly in microsomes. Liver injury caused by CCl_1 is used as a model for liver injury caused by xenobiotics [31–33]. In contrast to CCl_4, the galactose amine is water soluble, so the radical reactions induced by it take place in a watery compartment. Although similar to CCl_4, LPO plays some role in eliciting liver injury; it is not the primary cause of pathology in contrast to CCl_4. Liver injury caused by galactose amine can be considered a model of inflammatory liver injury (hepatitis caused by virus) [34]. It should be mentioned here that chronic administration of CCl_4, inducing injury by ROI production and resulting in the development of hepatomes in experimental animals, provides more evidence of the role of free radical reactions in carcinogenesis [31]. In clinical treatment of acute CCl_4 intoxication, the antioxidant therapy has proved useful, the intravenously administered N-acetyl cysteine substantially decreasing the occurrence of secondary hepatorenal injury [35].

Our investigations also proved the participation of free radical reaction in the hepatotoxic effect of CCl_4 and galactose amine. Following the administration of toxins, a markedly elevated MDA level and increased β-glucuronidase release were determined, which are the indicators of LPO injury. MTDQ could significantly decrease the hepatotoxic effect of CCl_4; laboratory parameters and morphological changes supported this observation. The water-soluble MTDQ-DS was also able to decrease significantly the hepatotoxic effect of galactose amine (MTDQ-DS, MTDQ) [36–40]. A similar protective effect of Catergen on the increased lysosomal enzyme release in chronic liver disease was detected in vitro [24, 41]. Furthermore, the liver-damaging effects of halothane (CF_3CHC_1Br), hydrazine (iproniazid, INH), hydralazine (Depressan), trichloroethylene, carbon disulfide, α-methyldopa (Dopegyt), acetaminophen (Phenacetin), nitrofurantoin, and paraquat are probably due to free radical reactions which can be presumed in liver injury caused by DDT, aflatoxin, and chlorpromazine (Hibernal) [8, 42, 43]. Of these materials, the mechanism of hepatotoxicity of halothane will be given in more detail due to its side effects in anesthesia. Halothane also requires metabolic activation for the development of hepatotoxicity. Its biotransformation takes place at the site of cytochrome P-450. Two metabolic pathways depending on oxygen tension exist, one of which is oxidative degradation, which results in a nontoxic, stable end product excreted in urine. The other pathway is a reductive route (inhibited by the rise in oxygen tension which is highly activated at low oxygen tension or under anaerobic conditions. During this process a radical ($CF_3 \cdot CHCl$) is generated which is able either to initiate LPO or to bind covalently to microsomal membrane compounds, including cytochrome P-450. LPO and covalent binding explain the decrease in the cytochrome P-450 level observed following halothane exposure in a hypoxic atmosphere. Hypoxic conditions may develop during anesthesia which increase the hepatotoxicity of halothane. According to epidemiological data, liver injuries induced by halothane are more frequent in obese patients. They might be due to the decreased oxygen supply of hepatocytes in obese patients, compared to those with normal weight. Hence in obese patients the damaging reductive metabolic pathway is dominant under hypoxic conditions [32, 44, 45]. N-Acetylcysteine in the treatment of acetaminophen intoxication and Legalon treatment in lethal *Amanita* poisoning have proved to be clinically effective and have saved the lives of such patients [1, 8].

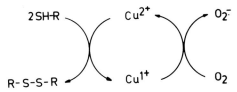

Fig. 3. Increased cellular and extra-cellular $O_2^{\cdot-}$ development observed in Wilson's disease. *SH-R*, sulfydryl; *R-S-S-R*, disulfide

Metabolic Liver Diseases

Wilson's Disease

In Wilson's disease an increased $O_2^{\cdot-}$ concentration in cells and extra-cellular space is found for several reasons:

1. The ceruloplasmin is an extracellular radical scavenger with mild SOD activity. A low ceruloplasm level typical of the disease decreases the antioxidant protection of the extracellular space.
2. Due to the ferroxidase character of ceruloplasmin, it catalyzes the ferrous → ferric iron oxidation, and its absence leads to ferrous iron auto-oxidation with production of $O_2^{\cdot-}$.
3. Finally, in ceruloplasmin deficiency free copper is deposited in tissues. SH groups can reduce Cu^{2+} to Cu^{1+}, which auto-oxidize and induce $O_2^{\cdot-}$ production (Fig. 3). The cirrhosis and extrapyramidal damage which developed in Wilson's disease support the possible free radical origin of cirrhosis and extrapyramidal clinical pictures of other origin. Moreover, apart from the above mechanisms, hemolytic episodes occasionally observed in Wilson's disease can be caused partly by the occurrence of copper in the circulation [46].

Hemochromatosis

There is evidence that the damage in secondary hemosiderosis (transfusion, β-Cooley's anemia) is caused by LPO; hence free radical reactions might play a role in the pathogenesis of the organic damage which occurs in hemochromatosis. This is evidenced by the presence of lipofuscin and ceroid in both Wilson's disease and hemochromatosis, which prove the presence of in vivo LPO. D-Penicillamine has a favorable effect not only as a metal chelator in Wilson's disease and hemochromatosis, but also probably due to the SOD activity of its copper complex [46–49].

Many liver protective drugs used in clinical practice have an antioxidant effect, e.g., vitamin E, lipoic acid, Legalon, D-penicillamine, and Catergen (previously used, but now not proposed due to its toxic side effects). In the treatment of toxic liver diseases the use of antioxidant compounds provides new opportunities; this is clearly demonstrated by the successful clinical use of N-acetylcysteine in CCl_4 and acetaminophen intoxication, and of Legalon in lethal *Amenita* poisoning. Our results and data from the literature seem to indicate the use of Legalon is promising in the treatment of widespread chronic alcoholic liver disease.

References

1. Fehér J, Vereckei A (1990) Májbetegségek. Diagnosztika és terápia. Medicina, Budapest
2. Bus JS, Gibson JE (1982/83) Mechanisms of superoxide radical mediated toxicity. J Toxicol Clin Toxicol 19:689
3. Heikkila RE, Cohen G (1973) 6-Hydroxydopamine: evidence for superoxide radical as an oxidative intermediate. Science 181:456
4. Trush MA, Mimnaugh EG, Ginsburg E, Gram TE (1982) Studies on the interaction of Bleomycin A_2 with rat lung microsomes II. Involvement of adventitious iron and reactive oxygen in Bleomycin mediated DNA chain breakage. J Pharmacol Exp Ther 221:159
5. Dormandy TL (1983) An approach to free radicals. Lancet ii:1010
6. Hornsby PJ, Crivello JF (1984) The role of lipid peroxidation and biological antioxidants in the function of the adrenal cortex. I A background review. Mol Cell Endocrinol 30:1
7. Fehér J, Vereckei A (1985) Szabadgyök reakciók jelentósége az orvostudományban. Biogal, Gyógyszergyár, Hungary
8. Fehér J, Csomós G, Vereckei A (1987) Free radical reactions in medicine. Springer, Berlin Heidelberg New York
9. Schraufstatter IV, Hyslop PA, Jackson J, Cochrane CC (1987) Oxidant injury of cells. Int J Tissue React 9:317
10. Younes M, Siegers CP (1984) Interrelation between lipid peroxidation and other hepatotoxic events. Biochem Pharmacol 33:2001
11. Ungemach FR (1987) Pathobiochemical mechanisms of hepatocellular damage following lipid peroxidation. Chem Phys Lipids 45:171
12. Cochrane CG, Schraufstatter IV, Hyslop PA, Jackson J (1987) Cellular and biochemical events in oxidant injury. In: Oxygen radicals and tissue injury. Proceedings of a Brook Lodge Symposium. Augusta, Michigan, USA, April 27–29, p 49
13. Ryle PR (1984) Free radicals, lipid peroxidation and ethanol hepatotoxicity. Lancet ii:461
14. Shaw S, Rubin KP, Lieber GS (1983) Depressed hepatic glutathione and increased diene conjugates in alcoholic liver disease. Evidence of lipid peroxidation. Dig Dis Sci 28:585
15. Ribiére C, Sinaceur S, Nordmann J, Nordmann R (1983) Liver superoxide dismutases and catalase ethanol inhalation and withdrawal. Pharmacol Biochem Behav 18 Suppl 1:263
16. Shaw S, Jayatilleke E, Ross WA, Gordon EF, Lieber CS (1981) Ethanol-induced lipid peroxidation potentiation by long-term alcohol feeding and attenuation by methionine. J Lab Clin Med 98:417
17. Pár A, Horváth T, Pakodi F, Zsoldos T, Kerekes E, Paál M, Kádal I, Beró T, Jávor T (1984) Hepatoprotective and immune moderator effects of antioxidants. In: Scientific session: Oxygen free radicals and tissue damage. Pécs, 10–11 January 1984
18. Oei HH, Bose SK, McCord JM (1985) Ethanol-induced oxidative stress: role of acetaldehyde in determining xanthine dehydro- genase/oxidase ratio. In: Fourth international conference on superoxide and superoxide dismutase. Rome, 1–6 September. Abstract p 89
19. Pár A, Jávor T (1984) Alternatives in hepatoprotection: cytoprotection-influences on monoxidase system-free radical scavengers. Acta Physiol Hung 64:409

20. Stege TE, Mischke BS, Cox GW, Daniels KA (1983) The role of free-radical inhibitors on acetaldehyde induced increases in lipid peroxidation. Fed Proc 47:513
21. Garcia-Bunuel L (1984) Lipid peroxidation in alcoholic myopathy and cardiomyopathy. Med Hypotheses 13:217
22. Emerit I, Braquet M, Congy F, Clavel JP (1984) Superoxyde dismutase érythrocytaire chez l'alcoolique chronique avec lésions hépatiques. Presse Med 13:1277
23. Goebel KM, Goebel FD, Schubotz R Schneider J (1979) Hemolytic implications of alcoholism in liver disease. J Lab Clin Med 94:123
24. Fehér J, Pollák Zs, Sréter L, Toncsev H, Cornides Á, Vereckei A (1984) Experimental models for the study of hepatoprotection. Acta Physiol Hung 64:401
25. Fehér J, Toncsev H, Fehér E, Kiss Á, Vasadi Á (1981) Lysosomal enzymes in sera and granulocytes of patients with chronic liver diseases. Int J Tissue React 3:31
26. Toncsev H, Fehér J, Fehér E (1980) Activity and release of beta-glucuronidase in granulocytes of patients with chronic liver diseases. Hepatology 9:22
27. Toncsev H, Fehér J, Fehér E, Vasadi Á (1981) Investigation of lysosomal enzymes in the serum and granulocytes of patients with chronic hepatic diseases. Magy Belorv Arch 34:293
28. Fehér J, Toncsev H, Sréter L, Cornides Á, Kiss Á (1983) Protection of hepatocyte membrane with antioxidant substances (cyanidanol-3, dihydroquinoline, etc.). In: Conference on the pathobiochemistry and pharmacology of liver. Mátrafüred, 23–24 September 1983
29. Fehér J, Láng I (1988) Wirkmechanismen der sogenannten Leberschutzmittel. Bayer Inter 4/88:35–39
30. Müzes Gy, Deák Gy, Láng I, Nékám K, Niederland V, Fehér J (1990) Silymarin (Legalon) kezelés hatása idült alkoholos májbetegek antioxidáns védórendszerére és a lipid peroxidációra (kettós vak protokoll). Orv Hetil 131, 16:863
31. McCay PB, King M, Lai EK, Poyer SL (1983) The effect of antioxidants on free radical production during in vivo metabolism of carbon tetrachloride. J Am Coll Toxicol 2:195
32. Reynolds ES, Treinen S, Moslen H (1980) Free-radical damage in liver. In: Pryor WA (ed) Free radicals in biology vol 4. Academic, New York, p 49
33. Slater TF (1981) Free radical scavengers. In: Conn HO (ed) International workshop on (+)cyanidanol-3 in disease of the liver. Royal Society of Medicine International Congress and Symposium series No. 47, p 11
34. Yoshikawa T, Kondo M (1982) Role of vitamin E in the prevention of hepatocellular damage: clinical and experimental approach. In: Lubin B, Machlin LJ (eds) Vitamin E: biochemical, hematological, and clinical aspects. Ann NY Acad Sci 393:198
35. Ruprah M, Mant TGK, Flanagan RJ (1985) Acute carbon tetrachloride poisoning in 19 patients: implications for diagnosis and treatment. Lancet i:1027
36. Fehér J, Bär-Pollák Zs, Sréter L, Fehér E, Toncsev H (1982) Biochemical markers in carbon tetrachloride- and galactosamine-induced acute liver injuries: the effects of dihydroquinoline-type antioxidants. Br J Exp Pathol 63:394
37. Fehér J, Pollák Zs, Sréter L, Fehér E (1983) Protection of hepatocyte membrane with antioxidant substances (cyanidanol-3, dihydroquinoline, etc.). In: Conference on the pathobiochemistry and pharmacology of liver. Mátrafüred, 23–24 September 1983
38. Sréter L, Kiss A, Cornides Á, Vereckei A, Toncsev H, Fehér J (1983) Inhibition of Doxorubicin-induced hepatic toxicity by a new dihydroquinoline type antioxidant. In: International symposium on recent advances in gastrointestinal cytoprotection. Pécs, 30 September–1 Oktober 1983. Acta Physiol Acad Sci Hung 64:431
39. Toncsev H, Bär-Pollák Zs, Kiss Á, Cornides Á Fehér J (1983) Protective effect of a new dihydroquinoline-type antioxidant on lysosomal damage in carbon tetrachloride induced free radical ractions. Kisérl Orvostud 34:16
40. Toncsev H, Pollák Zs, Kiss Á, Sréter L, Fehér J (1982) Acute carbon tetrachloride induced lysosomal membrane damage and the membrane protecting effect of a new dihydroquinoline-type antioxidant. Int J Tissue React 4:325

41. Fehér J, Toncsev H, Sréter L, Cornides Á, Vereckei A (1983) Protection of hepatocyte membrane with antioxidant substances (cyanidanol-3, dihydroquinoline, etc.). In: Conference on the pathobiochemistry and pharmacology of liver. Mátrafüred, 23–24 September 1983
42. Kappus H (1987) A survey of chemicals inducing lipid peroxidation. Chem Phys Lipids 45:105
43. Kappus H (1987) Oxidative stress in chemical toxicity. Arch Toxicol 60:144
44. Cusins MJ (1983) Halothane hepatitis: what's new? Drugs 19:1
45. DeGroot H, Noll T (1983) Halothane hepatotoxicity: relation between metabolic activation, hypoxia, covalent binding, lipid peroxidation and liver cell damage. Hepatology 3:601
46. Schneider EL, Reed JD (1985) Life extension. N Engl J Med 312:1159
47. Fridovich I (1978) The biology of oxygen radicals. Science 201:875
48. Heys AD, Dormandy TL (1981) Lipid peroxidation in iron-overloaded spleens. Clin Sci 60:295
49. Hubers H (1983) Iron overload: pathogenesis and treatment with chelating agents. Blut 47:61

Oxidative Stress in Chemical Toxicity

H. Kappus

The liver is one of the target organs for the toxicity of drugs and chemicals. Chemical toxicity is associated with either acute or chronic effects on the liver. The underlying mechanisms are not well understood. Besides activation of drugs to reactive intermediates which bind to several cellular macromolecules, a number of drugs can also lead to the activation of oxygen, resulting in highly reactive species of oxygen.

Mostly the respective drug is activated by enzymes to a free radical which then reacts with molecular oxygen. A prime example is the hepatotoxic agent carbon tetrachloride, which is active via the CCl_3 radical, to which molecular oxygen is bound. This reactive intermediate then initiates lipid peroxidation [1]. Many other drugs do not bind molecular oxygen, but rather transfer electrons to oxygen, resulting in the formation of a superoxide anion. This is the normal intermediary metabolite of oxygen generally formed during all oxidative processes in low amounts. Superoxide is destroyed by superoxide dismutase, leading to hydrogen peroxide and oxygen [2].

$$O_2 + e^- \rightarrow O_2^{\cdot -}$$
$$O_2^{\cdot -} + O_2^{\cdot -} + 2H^+ \rightarrow H_2O_2 + O_2$$

However, the interaction of superoxide and hydrogen peroxide leads to the hydroxyl radical, a very reactive oxygen species. This reaction is catalyzed by the so-called Haber-Weiss reaction.

$$O_2^{\cdot -} + H_2O_2 + H^+ \rightarrow HO^{\cdot} + O_2 + H_2O$$

It has been known for several years that the Haber-Weiss reaction is only possible if iron ions are present in catalytic amounts.

All reactive oxygen species formed in cells including singlet oxygen can oxidize a number of cellular constituents like lipids, proteins, and DNA (Fig. 1). The oxidation of lipids occurs in membranes, especially the intracellular membranes. During lipid peroxidation many different products, e.g., aldehydes, peroxides, and free radicals, are formed [3]. The direct effect of lipid peroxidation is the destruction of the membrane (Fig. 1). Secondary effects include the interaction of lipid peroxidation reaction products with other cellular components, e.g., the reaction of aldehydes with DNA, which can result in mutagenicity or carcinogenicity. On the

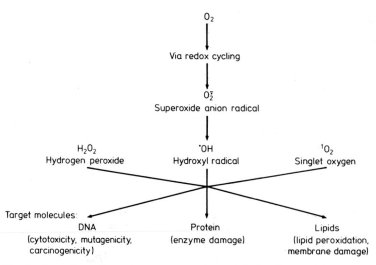

Fig. 1. Formation of reactive oxygen species and their effects on biomolecules. (From [1])

other hand, the oxidation of proteins leads to enzyme inactivation and thereby to toxicity (Fig. 1). The direct oxidation of DNA by highly reactive oxygen species gives oxidation products like hydroxylated bases, and the oxidation of the sugar moiety leads to strand breaks of DNA (see below).

This review will focus on a number of compounds which undergo redox cycling, a mechanism which continuously produces reactive oxygen species. In general, the drug is reduced by a cellular reductase, often cytochrome P-450 reductase or cytochrome b_5-reductase, whereby a free radical of the drug is formed [4]. In the presence of oxygen most of these free radicals can react with oxygen, resulting in superoxide anion radical formation. The activity of the reductases involved is relatively high in the liver, especially in the endoplasmic reticulum.

An example which is well studied is the anticancer drug Adriamycin (doxorubicin hydrochloride), which is reduced by NADPH-cytochrome P-450 reductase to the semiquinone radical (Fig. 2). As already pointed out, the semiquinone of Adriamycin can react with oxygen and can form super-oxide anion, which leads to other reactive oxygen species (see above). The mechanism of Adriamycin toxicity is, however, not very well understood, because the reductase also catalyzes splitting of the sugar moiety of Adriamycin, whereby the aglycone of Adriamycin is formed; but this re-duction also involves a free radical which might react with oxygen. Under certain conditions, Adriamycin is able to induce lipid peroxidation which is dependent on the presence of iron ions (see above). This lipid peroxidation step is probably induced via superoxide-catalyzed reduction of iron ions as shown in Fig. 3. That Adriamycin is not very toxic to the liver instead to the

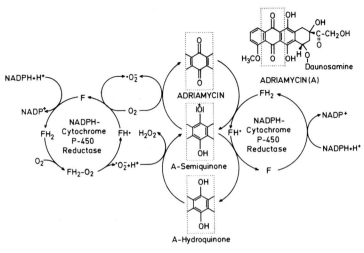

Fig. 2. Redox cycling of Adriamycin by NADPH-cytochrome P-450 reductase. (From [5])

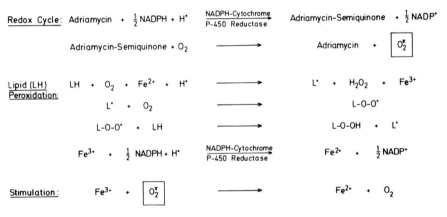

Fig. 3. Proposed mechanism of Adriamycin-induced lipid peroxidation in the presence of iron ions. (From [6])

heart is probably due to the high capacity of the liver with antioxidative mechanisms.

A further example of a quinonoid compound which can be reduced in a one-electron step is benzo[*a*]pyrene-quinone, a normally occurring metabolite of benzo[*a*]pyrene. This compound can be reduced by any reductase, forming the semiquinone, which reacts with oxygen leading to the superoxide anion (Fig. 4). It has been shown that this quinone of benzo[*a*]pyrene is mutagenic to bacteria which are only sensitive to oxidative stress. The mutagenicity was strongly dependent on the presence of NADPH-cytochrome P-450-reductase and oxygen, indicating that redox

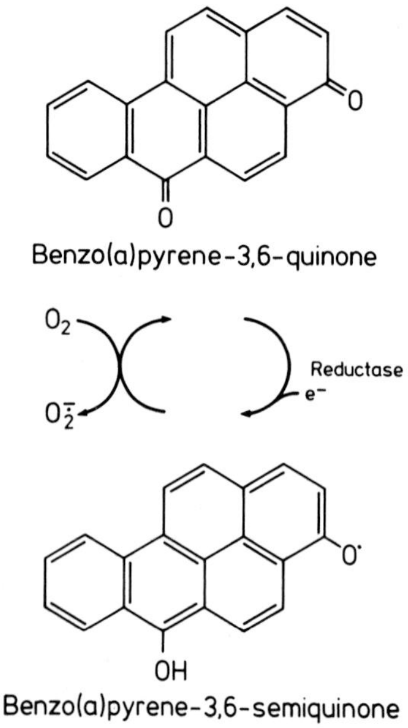

Benzo(a)pyrene-3,6-quinone

Reductase
e⁻

Benzo(a)pyrene-3,6-semiquinone

Fig. 4. Redox cycling of benzo[*a*]pyrene quinone

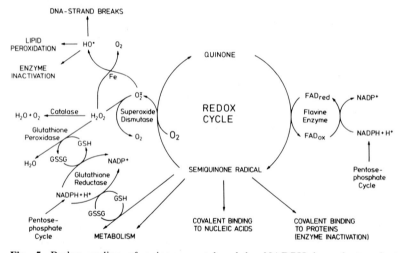

Fig. 5. Redox cycling of quinones catalyzed by NADPH-dependent reductases and reactive oxygen species formation, reaction, and inactivation. (From [4])

cycling is a prerequisite for the mutagenicity and that reactive oxygen species cause this effect [5].

Other quinonoid compounds which have been shown to undergo redox cycling are menadione, paraquat, alloxan, and a number of metabolites which are formed from non-quinonoid compounds. In general, all aromatic

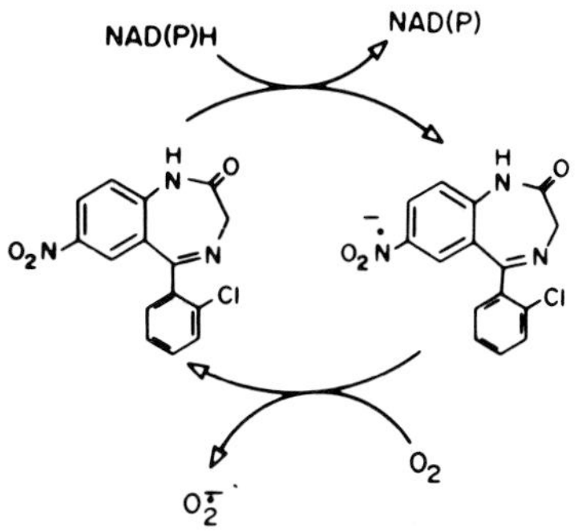

Fig. 6. Redox cycling of nitrazepam

chemicals which are oxidized to *ortho*- or *para*-quinones are able to redox cycle and form reactive oxygen species [2–5].

Not all of these compounds are hepatotoxic, a fact which is probably due to the very efficient equipment of the liver with antioxidative mechanisms, such as superoxide dismutase, catalase, and glutathione peroxidase. The various mechanisms of activation of quinones as related to their hepatotoxicity are shown in Fig. 5. Besides the reaction of the semiquinone radical with oxygen, this metabolite can also directly combine with proteins and nucleic acids, also leading to toxicity. Furthermore, if the semiquinone radical is further reduced to the hydroquinone, it can be conjugated with sulfate or glucuronide, a mechanism which can be regarded as a trapping reaction.

Besides quinones nitroaromatic compounds can also be reduced to free radicals and undergo redox cycling [6] as shown for example for nitrazepam (Fig. 6). Whether this also occurs in mammalian tissue is not very clear. However, some nitroaromatic chemotherapeutic agents have been shown to kill parasites by this redox cycling mechanism.

If the nitroaromatic compound is fully reduced, an aromatic amine is formed via the hydroxylamine. Hydroxylamines can also undergo reduction to free radicals which react with oxygen, resulting in the formation of the superoxide anion [6]. Via this reaction mechanism, aromatic amines can also undergo redox cycling. Not many aromatic amines or hydroxylamines are studied in relation to their liver toxicity and the formation of reactive oxygen species. But this mechanism is well established in erythrocytes where aromatic amines or hydroxylamines interact with hemoglobin, the underlying mechanism of toxicity being similar.

A rather peculiar situation exists with bleomycin, an antitumor drug which is mainly toxic to the lung but which as an iron complex can undergo redox cycling very efficiently with a number of liver enzymes [7, 8]. The oxidized iron complex of bleomycin is for example reduced by isolated NADPH-cytochrome P-450 reductase and the reduced iron complex binds and activates oxygen (Fig. 7). It has been shown that this reactive oxygen

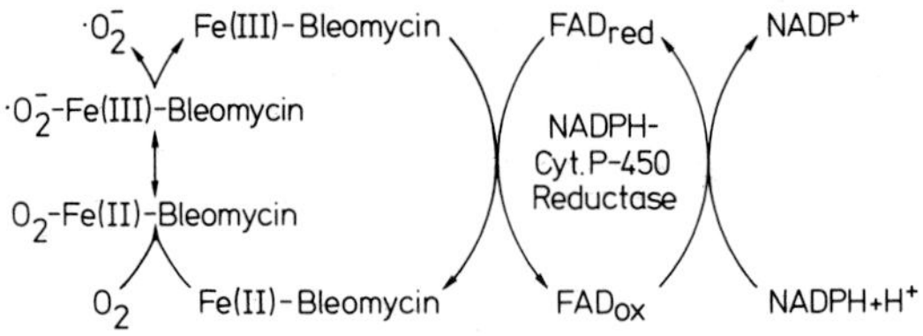

Fig. 7. Redox cycling of the bleomycin-iron complex catalyzed by NADPH-cytochrome P-450 reductase. (From [7])

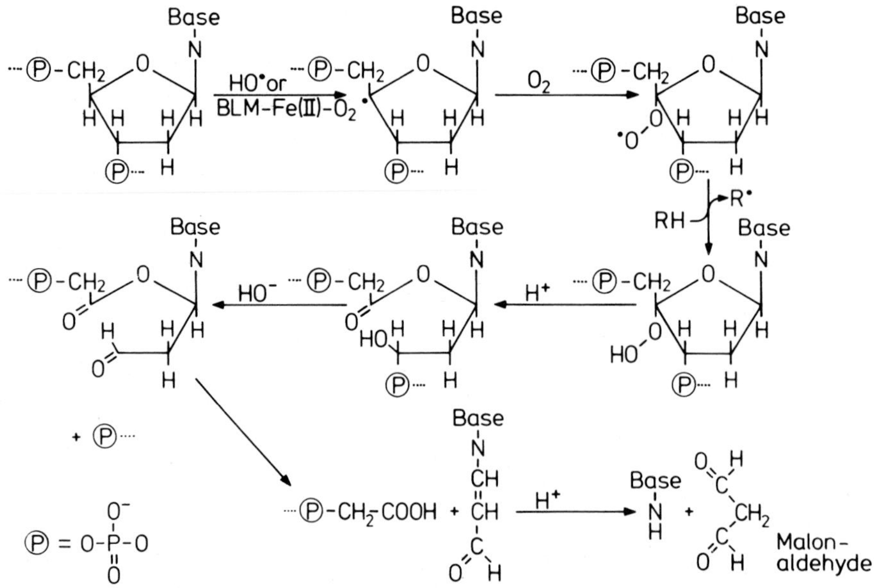

Fig. 8. Oxidative damage of the sugar moiety of DNA by "activated bleomycin" (BLM-Fe(II)-O_2), resulting in the formation of free bases, malondialdehyde, and DNA breaks. (From [9])

species formed during redox cycling of the bleomycin-iron complex leads to the oxidation of deoxyribose of DNA when it is present [9]. We and a number of other groups have shown that during this oxidation of DNA free bases and malondialdehyde are formed and strand breaks of DNA occur as shown in Fig. 8. The strand break formation might be responsible for the toxicity observed, especially in tumor cells. Furthermore, the reactive oxygen species could lead to the oxidation of DNA bases. We studied this in a reconstituted system containing bleomycin, iron ions, NADH-cytochrome b_5 reductase, cytochrome b_5, NADH and DNA, and analyzed 8-hydroxydeoxyguanosine formed in DNA. The result is shown in Fig. 9, which demonstrates that this reaction product is formed with increasing incubation time. The formation of 8-hydroxydeoxyguanosine was dependent on the presence of the enzyme and the cofactor NADH and bleomycin

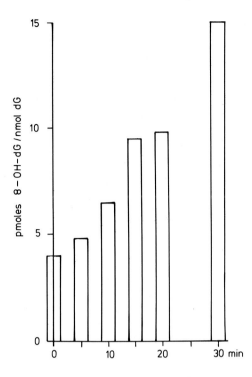

Fig. 9. Formation of 8-hydroxydeoxyguanosine (8-OH-dG) in DNA incubated with bleomycin, iron ions, NADH, cytochrome b_5 reductase, and cytochrome b_5 isolated from liver microsomes

as well as iron ions [10]. Parallel to the formation of hydroxylated bases of DNA, malondialdehyde is formed and NADH is consumed, indicating redox cycling of the bleomycin-iron complex, formation of reactive oxygen species, and occurrence of DNA strand breaks. It is likely that the formation of strand breaks during redox cycling of the bleomycin-iron complex causes acute toxicity, whereas the hydroxylation of the bases is related to the mutagenic and carcinogenic effects of bleomycin.

Reactive oxygen species are not only formed during redox cycling of drugs, but are also induced by depletion of protecting mechanisms against oxygen radicals. One example is the depletion of glutathione, which under certain conditions can lead to increased lipid peroxidation in the liver [1, 3]. However, this mechanism requires an additional stress which initiates the formation of reactive oxygen species, for example by uncoupling of the respiratory chain of mitochondria or by uncoupling of the hydroxylation reaction of cytochrome P-450. Depletion of vitamin E or other low molecular weight antioxidants can also lead to oxidative stress. But in this case other additional chemicals leading to oxidative stress are also required.

Under physiological conditions the amount of reactive oxygen species in cells is relatively low and the intermediates formed are processed very well by normal cells. Especially the liver has a very high capacity for antioxidant mechanisms, probably due to the fact that the liver has to process a number of chemicals, which can lead to oxidative stress. But under disease conditions or in the presence of various chemicals these protective mechanisms of the liver can be overwhelmed, whereby reactive oxygen species are no longer trapped and can cause toxicity such as membrane damage, enzyme inactivation, and DNA breaks.

In conclusion, oxidative stress is a common mechanism by which chemical toxicity can occur in the liver. The enzymes involved are in the case of redox cycling drugs reductases which are present at levels which produce high activities in the liver. Redox cycling of chemicals could be increased in the liver by the low oxygen concentration present, which favors the reduction of drugs, the oxygen concentration still being high enough to produce reactive oxygen species in high amounts. But other mechanisms besides redox cycling also have to be taken into account when considering oxidative stress and chemical hepatotoxicity.

References

1. Kappus H (1987) A survey of chemicals inducing lipid peroxidation in biological systems. Chem Phys Lipids 45:105–115
2. Kappus H, Sies H (1981) Toxic drug effects associated with oxygen metabolism: redox cycling and lipid peroxidation. Experientia 37:1233–1241
3. Kappus H (1985) Lipid peroxidation: mechanisms, analysis, enzymology and biological relevance. In: Sies H (ed) Oxidative stress. Academic, London, pp 273–310
4. Kappus H (1986) Overview of enzyme systems involved in bioreduction of drugs and in redox cycling. Biochem Pharmacol 35:1–6
5. Kappus H (1987) Oxidative stress in chemical toxicity. Arch Toxicol 60:144–149
6. Kappus H, Muliawan H, Scheulen ME (1984) The role of iron in lipid peroxidation induced by adriamycin during redox cycling in liver microsomes. In: Bors W, Saran M, Tait D (eds) Oxygen radicals in chemistry and biology. de Gruyter, Berlin, pp 359–362
7. Scheulen ME, Kappus H, Thyssen D, Schmidt CG (1981) Redox cycling of Fe(III)-bleomycin by NADPH-cytochrome P-450 reductase. Biochem Pharmacol 30:3385–3388
8. Mahmutoglu I, Kappus H (1988) Redox cycling of bleomycin-Fe(III) and DNA degradation by isolated NADH-cytochrome b_5 reductase: involvement of cytochrome b_5. Mol Pharmacol 34:578–583
9. Mahmutoglu I, Scheulen ME, Kappus H (1987) Oxygen radical formation and DNA damage due to enzymatic reduction of bleomycin-Fe(III). Arch Toxicol 60:150–153
10. Kappus H, Bothe D, Mahmutoglu I (1990) The role of reactive oxygen species in the antitumor activity of bleomycin. Free Radic Res Commun 11:261–266

Relationships Between Free Radical Reactions and the Function of the Cytochrome P-450 System

V.E. Kagan, L. Packer, and E.A. Serbinova

Introduction

The metabolically uncoupled oxidation of polyunsaturated fatty acid residues of membrane phospholipids, proteins, and DNA is fundamental to the development of numerous pathologies and toxic effects [1]. During the normal course of metabolism, approximately 98% of molecular oxygen undergoes complete reduction to water via the "oxidase" pathway [2]. The remainder is converted to partially reduced products (superoxide anion radical, hydrogen peroxide, hydroxyl radical) which are capable of initiating free radical mediated oxidative damage. These reactive oxygen species may arise from different sources, the most important of which are mitochondrial respiration, phagocytic and bactericidal function of leukocytes and macrophages, dioxygenase activities (e.g., cyclooxygenase and lipoxygenase), and cytochrome P-450-supported monooxygenations [3]. A characteristic feature of free radical reactions is their ability to initiate and propagate chain reactions, resulting in an amplification and involvement of the bulk of membrane polyunsaturated lipids. Transition metals can act as potent catalysts of free radical reactions both by enhancing formation of highly reactive hydroxyl radicals in the aqueous phase in the course of the initiation stage and by splitting down hydroperoxides with subsequent formation of reactive alkoxyl and peroxyl radicals in the membranous phase at the propagation stage [4].

Rat liver microsomes generate a variety of potent oxidants during NADPH-dependent electron transport which under appropriate conditions (in the presence of a suitable transition metal chelate as a catalyst) may promote lipid peroxidation in microsomal membranes [5].

Free radicals initiate lipid peroxidation of microsomal membranes by rapidly forming lipid hydroperoxides (LOOH) from endogenous polyunsaturated fatty acid residues of phospholipids [6]. Lipid hydroperoxides may in turn propagate lipid peroxidation via the cytochrome P-450-dependent mechanism. Indeed cytochrome P-450 catalyzes the oxidative cleavage of lipid hydroperoxides to a pool of alkoxyl (LO) and peroxyl (LOO) radicals which induce an additional formation of lipid hydroperoxides. Propagation and termination of lipid peroxidation in the cell are accompanied by a concomitant accumulation of numerous lipid peroxidation

Table 1. Effects of reticulocyte 15-lipoxygenase (15-RLOX)-catalyzed accumulation of phospholipid hydroperoxides in rat liver microsomes on aniline hydroxylase (AH) and ethylmorphine demethylase (EMD) activities

Additions	Activities	
	EMD (nmol HCHO/mg/min)	AH (nmol PAPh/mg/min)
Microsomes	0	0
+ NADPH-generating system[a]	3.50 ± 0.3	0.40 ± 0.05
+ 15-RLOX[b]	1.32 ± 0.2	0.03 ± 0.001
+ 15-RLOX + NADPH-generating system	40.6 ± 0.2	0.76 ± 0.06

PAPh, p-aminophenol.
[a] NADPH-generating system consisted of NADP, isocitrate, and isocitrate dehydrogenase;
[b] Addition of 15-RLOX resulted in oxidation of 15 mol% of microsomal membrane phospholipids.

products possessing different effects on biomembranes and membrane-bound enzymes [3, 7]. However, these products originate from hydroperoxides, the primary molecular lipid peroxidation products, which can be reduced to corresponding hydroxy compounds by peroxidases [8], thus preventing the formation of various scission products.

Results and Discussion

Lipid peroxidation is considered to be an efficient triggering mechanism of the disassembly of microsomal membranes and cytochrome P-450. An inverse correlation exists between the steady-state concentrations of lipid peroxidation products and cytochrome P-450 content (activities of cytochrome P-450-supported reactions) in liver endoplasmic reticulum membranes [9]. Thus lipid peroxidation may activate the consequence of reactions (including activation of endogenous phospholipases and proteases) resulting in disassembly of cytochrome P-450 [10].

One of the principle points is the reversibility of changes occurring in microsomal membrane in the course of lipid peroxidation. It is obvious that at the initial stage when lipid hydroperoxides are accumulated as the primary molecular products of peroxidation oxidative modification may be reversed due to peroxidase-catalyzed reduction of hydroperoxides to hydroxy compounds. The reversibility at this stage is still possible if lipid hydroperoxides themselves do not destroy cytochrome P-450.

The comparison of kinetic curves of cytochrome P-450 destruction in liver microsomal membranes with the curves of lipid hydroperoxides and secondary lipid peroxidation products [thiobarbituric acid (TBA)-reacting carbonyl compounds] in vitro shows that there is no correlation between

Table 2. Spearman's correlation coefficients for cytochrome P-450 content and lipid peroxidation products in rat liver microsomes in the course of $(Fe^{2+} + ascorbate)$-induced lipid peroxidation

	Hydroperoxides	TBA-reactive substances
Cytochrome P-450	−0.2857	−1.000
p value	0.5350	0.000

degradation of the hemoprotein and accumulation of hydroperoxides (Table 1). The time course of lipid hydroperoxides passes through a maximum, whereas cytochrome P-450 declines following monotonous kinetics as does the accumulation of TBA-reactive substances [11]. Thus it seems unlikely that primary lipid peroxidation products, hydroperoxides, are directly involved in cytochrome P-450 disassembly. This conclusion was confirmed by our direct measurements where only phospholipid hydroperoxides (but not the mixture of different lipid peroxidation products) were tested as modifiers of cytochrome P-450 [11]. In these experiments we used reticulocyte lipoxygenase, which can attack polyunsaturated phospholipids to form hydroperoxides of polyunsaturated fatty acid residues as a predominant product [12]. We found that the increasing concentrations of phospholipid hydroperoxides did not cause the destruction of cytochrome P-450 (or its conversion into catalytically inactive form P-420). Moreover, in the presence of phospholipid hydroperoxides NADPH-dependent hydroxylation reactions were stimulated (Table 2).

Thus we conclude that lipid peroxidation does not destruct cytochrome P-450 until secondary peroxidation products are accumulated in sufficient amounts. Control of lipid peroxidation due to GSH-peroxidase-dependent reactions is one of the mechanisms to prevent scission of hydroperoxides. The other possibility is the interaction of lipid peroxyl radicals (LOO) with antioxidant molecules, like vitamin E (α-tocopherol, TOC-OH):

$$LOO\cdot + TOC\text{-}OH \xrightarrow{\text{Free radical reductase}} LOOH + TOCO\cdot$$

However, this reaction results in an exhaustion of membrane vitamin E pools. It was suggested that special enzyme system(s), free radical reductase(s), may operate to recycle antioxidants from their phenoxyl radicals. No direct experimental evidence supporting this hypothesis was obtained until recently [13, 14].

A reasonable question: why is vitamin E that crucially important for the overall antioxidant protection of membranes? There are at least two answers to this question: (1) Vitamin E is considered to be the major if not the only chain-breaking antioxidant of membranes [15]. (2) Vitamin E functions in the membranes as a center for antioxidation which harvests (collects) the antioxidant power not only from other lipid-soluble antioxidants (like

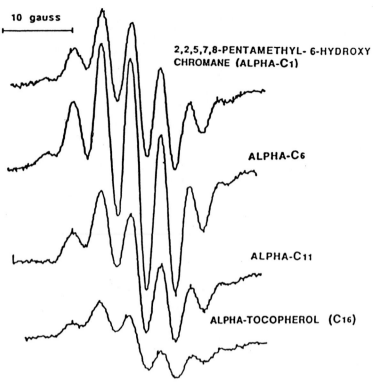

10 gauss

2,2,5,7,8-PENTAMETHYL- 6-HYDROXY
CHROMANE (ALPHA-C1)

ALPHA-C6

ALPHA-C11

ALPHA-TOCOPHEROL (C16)

Fig. 1. Electron spin resonance spectra of chromanoxyl radicals generated by the (lipoxygenase + linolenic acid) oxidation system in the presence of rat liver microsomes. Incubation medium contained: microsomes, 27 mg protein/ml; chromanols, 8 mM; linolenic acid, 14 mM; lipoxygenase 90 U/μl

ubiquinols) but also from reductants in the cytosolic phase (ascorbate, glutathione, etc.) (see our results below). Thus vitamin E is not only the major, but also the most efficient, membrane antioxidant. That is why the maintenance of steady-state concentration of vitamin E in the membranes may be crucial for their protection against antioxidant invasions.

We have developed simple and convenient methods to generate phenoxyl radicals from natural and synthetic antioxidants based on their: (1) enzymatic oxidation by (lipoxygenase + polyunsaturated fatty acid, e.g., arachidonic, linolenic) system or (2) nonenzymatic oxidation by the diazo-initiator of peroxyl radicals: 2,2'-azo-bis-(2,4-dimethylvaleronitrile), AMVN [16, 17]. Lipoxygenase generates peroxyl radicals of polyunsaturated fatty acids. In both systems peroxyl radicals interact with vitamin E to generate chromanoxyl radicals. Thus both systems very nicely imitate interaction of tocopherol with peroxyl radicals in the course of lipid peroxidation (Fig. 1).

Using these methods we were able to demonstrate that ascorbate is efficient in regeneration of chromanoxyl radicals of tocopherol and its

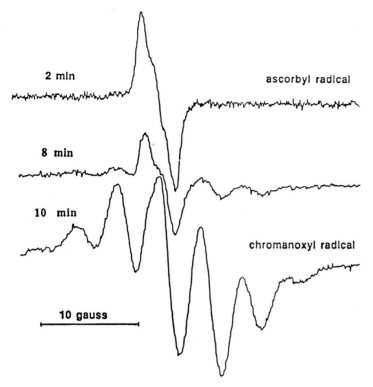

Fig. 2. Electron spin resonance spectra of chromanoxyl radicals generated from α-tocopherol in the presence of ascorbate. Concentration of ascorbate was 2.5 mM. Other conditions as in Fig. 1

homologues not only in liposomes but also in natural membranes, i.e., liver microsomes and submitochondrial particles (Fig. 2). Enzyme-dependent mechanisms which prevent accumulation of chromanoxyl radicals derived from vitamin E homologues were characterized in these membranes. NADPH or NADH in microsomes as well as NADH or succinate in mitochondria prevented accumulation of chromanoxyl radicals until these substrates were fully consumed (Fig. 3). Thus we concluded that rat liver microsomes and mitochondria have both enzymatic electron transport-dependent and nonenzymatic mechanisms for reducing chromanoxyl radicals.

There is now a substantial amount of experimental data which support the notion that CoQn, besides its well-recognized role as a redox component of the electron transport system, may function in its reduced or semireduced form as antioxidants in various biological membranes. However, the specific molecular mechanisms responsible for antioxidant activity of ubiquinones are still debated. In a direct way, reduced forms of CoQn may behave as

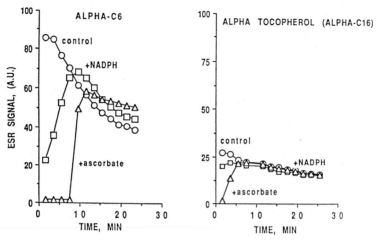

Fig. 3. Time course of chromanoxyl radicals from α-tocopherol and its homologue generated by the (lipoxygenase + linolenic acid) oxidation system in the presence of rat liver microsomes and their recycling by NADPH and ascorbate. Concentration of NADPH was 5 mM

other free radical scavengers, donating an H atom to peroxyl and alkoxyl radicals. Another possibility of indirect antioxidant function of CoQn was suggested by Mellors and Tappel [18], who put forward a hypothesis on vitamin E recycling by ubiquinones.

To obtain evidence of direct or indirect mechanism of antioxidant action of ubiquinols (ubiquinones) we: (1) compared the antioxidant efficiencies of tocopherol and ubiquinol (ubiquinone) homologues in rat liver microsomal and mitochondrial membranes under conditions of (Fe^{2+} + ascorbate)-induced lipid peroxidation and (2) studied the recycling of tocopherol and its homologues from the chromanoxyl radicals by ubiquinone-dependent electron transport in microsomes and mitochondria. We were able to demonstrate that tocopherols are much stronger membrane antioxidants than naturally occurring ubiqinols (ubiquinones) [19]. Thus direct radical scavenging effects of ubiquinols (ubiquinones) might be negligible in the presence of comparable or higher concentrations of tocopherols. In support of this, our ESR findings show that ubiquinones synergistically enhance enzymic NADH- and NADPH-dependent recycling of tocopherols by electron transport in mitochondria and microsomes (Fig. 4). Thus we conclude that antioxidant effects of ubiquinols (ubiquinones) are not due to their direct radical scavenging reactivity, but rather result from their ability to stimulate more efficient recycling of tocopherols interacting with electron-transport enzymes.

Thus chromanoxyl radicals of vitamin E can be reduced by ascorbate, NADPH-, NADH-, and succinate-dependent electron transport in microsomes and mitochondria, whereas glutathione, dihydrolipoate, and ubiquinols

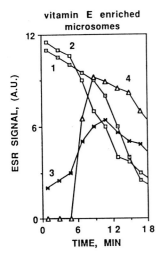

Fig. 4. Time course of ESR signal of chromanoxyl radicals in rat liver microsomes isolated from vitamin E supplemented animals (21.5 nmol α-tocopherol/mg protein). *1*, control; *2*, + CoQ1 (40.0 mnol/mg protein); *3*, + NADPH (7.5 mM); *4*, + CoQ1 + NADPH

synergistically enhance vitamin E regeneration. We conclude that vitamin E molecules possess a unique ability to act as membrane free radical harvesting centers which collect their antioxidant power from other intramembrane and cytosolic reductants. In this way vitamin E can modulate the activity of microsomal cytochrome P-450-dependent reactions.

Acknowledgements. This study was supported by NIH grant CA47597 and by a grant from the Bulgarian Academy of Sciences.

References

1. Horton AA, Fairhurst S (1987) Lipid peroxidation and mechanisms of toxicity. CRC Crit Rev Toxicol 18:27–79
2. Jamieson D (1989) Oxygen toxicity and reactive oxygen metabolites in mammals. Free Radic Biol Med 7:87–108
3. Halliwell B, Gutteridge JMC (1985) Free radicals in biology and medicine. Clarendon, London
4. Miller DM, Buettner GR, Aust SD (1990) Transition metals as catalysts of autoxidation reactions. Free Radic Biol Med 8:95–108
5. Sevanian A, Nordenbrand K, Kim E, Ernster L, Hochstein P (1990) Microsomal lipid peroxidation: the role of NADPH-cytochrome P-450 reductase and cytochrome P-450. Free Radic Biol Med 8:145–152
6. Hochstein P, Ernster L (1963) ADP-activated lipid peroxidation coupled to the TPNH-oxidase system in microsomes. Biochem Biophys Res Commun 12:388–394
7. Wills ED (1971) Effect of lipid peroxidation on membrane-bound enzymes of endoplasmic reticulum. Biochem J 123:983–991
8. Ursini F, Maiorino M, Gregolin C (1986) Phospholipid hydroperoxide glutathione peroxidase. Int J Tissue React 8:99–103
9. Kagan VE, Kotelevtzev SV, Koslov YuP (1974) Role of enzymic lipid peroxidation in disassembly of liver endoplasmic reticulum membranes. Proc Natl Acad Sci USSR 217:213–216

10. Davies KJA, Delsignore MI (1987) Protein damage and degradation of oxygen radicals. J Biol Chem 262:9895–9901
11. Serbinova EA, Ivanova S, Kitanova S, Packer L, Kagan V (1990) Cytochrome P-450 under conditions of oxidative stress: role of antioxidant recycling in the protection mechanisms. In: McCabe M, Goldstick T, Maquire D (eds) Advances in biology and medicine (18th international meeting of oxygen transport to tissue). Plenum, New York
12. Rapoport SM, Schewe T, Weisner R, Halangk W, Lunwig P, Janicke-Hohne M, Tannert C, Hiebsch C, Klatt D (1979) The lipoxygenase of reticulocytes. Purification, characterization and biological dynamics of lipoxygenase; its identity with the respiratory inhibitors of the reticulocyte. Eur J Biochem 96:545–565
13. McCay PB (1985) Vitamin E: interactions with free radicals and ascorbate. Annu Rev Nutr 5:323–340
14. Bast A, Haenen GR (1988) Interplay between lipoic acid and glutathione in the protection against microsomal lipid peroxidation. Biochim Biophys Acta 963:558–561
15. Burton GW, Ingold KU (1986) Vitamin E: application of the principles of physical organic chemistry to the exploration of its structure and function. Acc Chem Res 19:194–201
16. Packer L, Maquire J, Melhorn R, Serbinova E, Kagan V (1989) Mitochondria and microsomal membranes have a free radical reductase activity that prevents chromanoxyl radical accumulation. Biochem Biophys Res Commun 159:229–235
17. Kagan VE, Serbinova EA, Packer L (1990) Recycling and antioxidant activity of tocopherol homologs of differing hydrocarbon chain lengths in liver microsomes. Arch Biochem Biophys 282:221–225
18. Mellors A, Tappel AL (1967) Quinones and quinols as inhibitors of lipid peroxidation. Lipids 4:282–284
19. Kagan V, Serbinova E, Packer L (1990) Antioxidant effects of ubiquinones in microsomes and mitochondria are mediated by tocopherol recycling. Biochem Biophys Res Commun 169:851–857

Regulation Mechanisms of Xenobiotics Metabolizing Liver Cytochrome P-450 and Toxicological Implications

H. REIN, O. RISTAU, J. BLANCK, and K. RUCKPAUL

Cytochromes P-450 exhibit a key function in the biotransformation of xenobiotics [13]. Depending on the chemical nature of the foreign compounds, either a detoxification or a toxic activation is effected by the enzymes. The enzyme action is connected with the reductive splitting of molecular oxygen and the insertion of an oxygen atom into the substrate, thus converting a hydrophobic compound into a more polar product. The excretion of such a transformed compound is facilitated. The insertion of a reactive group into the molecule, however, induces the possibility of its covalent binding to proteins and nucleic acids resulting in toxic effects, even in mutagenesis and carcinogenesis. Besides this oxidative metabolism, cytochrome P-450 also catalyzes the reductive transformation of foreign compounds. This type of pathway mostly results in toxification since reactive free radicals are formed.

Polyhalogenated alkanes like carbon tetrachloride and the anesthetic halothane are examples of a reductive pathway, and the formed free radicals by dehalogenation may be directly involved in lipid peroxidation. Cytochrome P-450 enzymes, moreover, beside the bioactivation of xenobiotics exhibit an oxidase activity which results in the formation of activated oxygen species, e.g., superoxide radicals and hydrogen peroxide, which are also of toxicological importance, by initiating lipid peroxidation and their attack on proteins and nucleic acids.

The liver is the main locus of the cytochrome P-450 enzymes; however, lower concentrations are present in almost all tissues with the exception of skeletal muscle and erythrocytes. Cytochromes P-450 are monomeric hemoproteins which exhibit a molecular weight between 45000 and 60000. They exist in several isozyme forms which are characterized by specificities in substrate metabolism.

More than 20 different rat liver cytochromes P-450 have been purified and characterized; also several forms from rabbit and mouse liver are known, and from human liver at present 14 isozymes have been identified [15]. Today the primary structure of about 100 cytochromes P-450 are known, mostly based on cDNA sequence analysis [7]. Mammalian cytochromes P-450 are hydrophobic membrane proteins which are tightly associated with intracellular membranes. Those forms metabolizing xenobiotics are localized in the endoplasmic reticulum; some steroid hydroxylating enzymes are also present in the inner membrane of mitochondria. The activation of molecular oxygen by cytochrome P-450 requires electrons from the donor NADPH-cytochrome P-450 reductase, which is also membrane-bound. In the case of the endoplasmic cytochromes P-450 the competent reductase is a flavine adenine dinucleotide (FAD), flavin mononucleotide (FMN) containing flavoprotein which donates single electrons to the monooxygenase. Because of the membrane-bound state of both proteins the intermolecular electron transfer between the electron donor reductase and the electron acceptor cytochrome P-450 depends on the properties of the membrane phospholipids.

The multicomponent character of the cytochrome P-450 enzyme system (monooxygenase, reductase, phospholipid), its membrane-bound state, and the existence of several isozymes with different substrate specificities determines activity control at three different levels:

1. by intramolecular substrate-induced conformational changes of cytochrome P-450,
2. by the intermolecular cytochrome P-450/reductase interaction and its modulation by phospholipids,
3. by the genetically determined isozyme pattern and quantitative alterations by inducers.

Further activity control is exerted by the oxidase activity of cytochromes P-450 (uncoupling):

4. during the process of dioxygen activation reduced oxygen species, i.e., superoxide radicals, hydrogen peroxide, and water, are released, thus diminishing substrate oxidation. These pathways are isozyme and substrate specifically regulated.

The regulation of the enzyme activity of cytochrome P-450 at the molecular level (v. 1.) – especially the phenobarbital-inducible isozyme cytochrome P-450 2B4 – could be evidenced to depend on an equilibrium between two protein conformational states both characterized by different optical and magnetic properties [10]. The low spin state is characterized by a total spin of $S = 1/2$ and a position of the Soret band at 418 nm; the high spin state exhibits the Soret band at 387 nm and a spin value of $S = 5/2$. The equilibrium between both conformers is shifted by both substrates and temperature. In general, substrates shift the equilibrium toward the high spin state; the magnitude of this shift exhibits substrate specificity. The latter is based on different affinities of the two spin conformers toward the substrate, e.g., benzphetamine exhibits an approximately fivefold increased affinity to the high spin conformer of cytochrome P-450 2B4 as compared to the low spin conformer [11]. Substrate-induced spin transitions are correlated with changes of the strong negative redox potential of cytochromes P-450 in the absence of substrate (-359 mV for the isozyme 2B4) toward more positive values (-317 mV for benzphetamine).

Extrapolation of that correlation results in a redox potential of the high spin conformer of $-296\,mV$ [14]. The increased redox potential of the high spin cytochrome P-450 implies a favorable reduction of this conformer. Therefore a correlation between the high spin fraction (P-450$_{hs}$/P-450$_{hs}$ + P-450$_{ls}$) of the substrate-dependent spin equilibrium and the reduction rate constant has been evidenced [2].

The control of the electron transfer between the NADPH-dependent cytochrome P-450 reductase and cytochrome P-450 in the second reaction step of the catalytic cycle finally regulates dioxygen activation and oxygen transfer into the substrate. This coupling between substrate-induced high spin shift and substrate turnover was experimentally proved for mammalian liver microsomes [2] and a reconstituted cytochrome P-450 system and a series of benzphetamine derivatives [16].

The regulation of cytochrome P-450 enzyme activity at the membranous level (v. 2.) has to consider the direct interaction between cytochrome P-450 and reductase in the electron transfer between both proteins. Phospholipids as the third component of the cytochrome P-450 enzyme system mediate this interaction. It was shown that lipids with negatively charged head groups favor the complex formation between the electron donor and electron acceptor protein (RP) which is prerequisite for an effective electron exchange rate ($v = k \cdot RP$) [3]. The lipid-specific donor/acceptor dissociation constant K_{RP} was shown to differ by about one order of magnitude in favor of negatively charged lipids: $K_{RP} = 0.051\,\mu M$ (phosphatidylserine/oleoylphosphatidylethanolamine), $K_{RP} = 0.47\,\mu M$ (dioleoylphosphatidylcholine) in a reconstituted cytochrome P-450 (2B4) enzyme system.

The charge dependence of the lipid control in the cytochrome P-450 reduction suggests charge-dependent electrostatic interactions between the electron-exchanging proteins. Such interactions were analyzed by chemical modification of the N-terminal amino group and a functionally linked lysine of cytochrome P-450 (2B4) which results in restricted electron transfer and catalytic activity [1].

With respect to the isozyme pattern (v. 3.) the cytochrome P-450 of the liver-metabolizing xenobiotics exists as an isozyme family with members of distinct but overlapping substrate specificity. Each isozyme represents an individual gene product evidenced by amino acid and/or cDNA nucleotide sequencing. Some of these isozymes are constitutive; some of them can be induced by various inducing agents to a different extent. Depending on the specific pattern and the concentration of the individual isozymes, the biotransformation of xenobiotics differs both in rate and type of products. Inducers therefore are able to influence in a characteristic way detoxifying as well as toxifying conversions of substrates. The toxicological consequences of cytochrome P-450 induction by a lot of different chemical agents include cytochrome P-450 mediated mutagenesis and carcinogenesis. An example of the formation of different reactive intermediates in dependence on the induction represents the metabolism of benz[a]pyrene: The ratio of the formation of the primary K-region epoxide (4,5-epoxide) to the non-K-region epoxide in the 7,8-position is governed by the ratio of cytochrome P-450 2B4 to cytochrome P-450 2B4. Thus the induction of cytochrome P-450 1B2 by phenobarbital leads to a detoxification because the 4,5-epoxide is excreted via conjugation. Induction of cytochrome P-450 1B2 by 3-methylcholanthrene, however, results mainly in the formation of the 7,8-epoxide, which is converted to the corresponding diol by the epoxide hydrase. This diol is transformed by a further cytochrome P-450 1B2 dependent reaction to the ultimate carcinogen benzo[a]pyrene-7,8-diol-9, 10-epoxide [5].

Different pathways in dependence on the isozyme pattern of cytochrome P-450 are further observed at the hydroxylation of 2-acetylaminofluorene [17]. By use of different isozymes in reconstituted cytochrome P-450 enzyme systems, regiospecific hydroxylations of this compound were obtained. N-hydroxylation as catalyzed by cytochrome P-450 form 4 results in the ultimate carcinogen. Hydroxylation of the ring system (in position 7), on the other hand, is catalyzed by the isozymes form 3 and form 6 (form 3 is constitutive, form 6 is induced by TCDD in neonatal rabbit liver) and results in the detoxification of 2-acetylaminofluorene.

As to uncoupling (v. 4.) the correlation between the substrate-induced high spin shift and the substrate turnover (v. 1.) is remarkable because of the absence of disturbances by substrate-specific oxidase activities. That implies the latter to underly distinct correlations with the spin shift, too. Of special interest is that question with respect to complete uncouplers, e.g., perfluorocarbons, which activate cytochrome P-450 by initiating a high spin shift and thereby increase the NADPH oxidation, but cannot be hydroxylated because of their chemically inert C-F bonds instead of the C-H ones [4]. In the presence of these pseudosubstrates NADPH and dioxygen consumption is exclusively used for the oxidase pathways. Quantitation of NADPH oxidation, dioxygen consumption, and hydrogen peroxide formation reveals that obviously in this case beside the hydrogen peroxide formation also water is produced in a four-electron reduction process [8].

Our recent studies of oxygen consumption and hydrogen peroxide formation by the use of a reconstituted cytochrome P-450 enzyme system (isozyme 2B4) and several tertiary amines as substrates exhibit the following results [4]: Along with the substrate-induced high spin shift and substrate turnover (v. 1.) oxygen consumption also increases, and hydrogen peroxide formation remains constant. Therefore, within this substrate series the former correlation is not perturbed by the oxidase activity. Similar findings have been obtained with rabbit liver microsomes with respect to spin shift-dependent substrate conversion, hydrogen peroxide formation, and NADPH consumption. Moreover, by relating the enzyme activities to NADPH consumption, a four-electron-dependent water formation could be shown to decrease with the spin shift. Obviously water production and substrate turnover are complementary processes. The molecular basis for

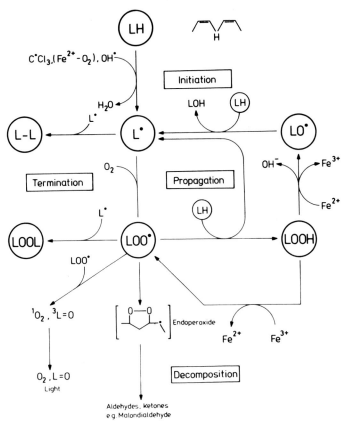

Fig. 1. Proposed pathways of lipid (L) peroxidation. (Modified scheme from [9])

such a correlation could be the fitting of the substrate to the active site. A tightly accommodated substrate results in an intensive high spin shift and an improved oxygen insertion into the substrate. The specific properties of the substrate are important for both effects.

Toxicological Implications

The toxicological consequences of the hydrogen peroxide formation by cytochrome P-450 are exhibited in substrate- and isozyme-dependent lipid peroxidation and destruction of cytochrome P-450 [12]. It is assumed that lipid peroxidation by hydrogen peroxide occurs via the very reactive hydroxyl radicals produced by the Haber-Weiss and/or the Fenton reaction.

The proposed mechanism of lipid peroxidation is shown in Fig. 1. It is assumed that hydrogen abstraction from methylene carbons of unsaturated fatty acids by partially reduced oxygen species, e.g., hydroxyl radicals, is the initiation step of lipid peroxidation. The resultant lipid radical reacts with molecular oxygen by producing a fatty acid peroxyl radical. This reactive intermediate abstracts a hydrogen atom from another fatty acid, resulting in a further peroxyl radical after reaction with dioxygen. Cytochrome P-450 may have a function in this self-propagating process by catalyzing homolytic or heterolytic splitting of lipid hydroperoxides to produce either free radicals or activated oxygen species. Lipid peroxidation generates reactive intermediates such as 4-hydroxy alkenal. These and other aldehydes formed by decomposition of lipids have been identified as major cytotoxic carbonyl compounds. Moreover, activated oxygen species, e.g., singlet oxygen and compounds such as products of lipid peroxidation, could contribute to the membrane destruction [9]. As to the isozyme specificity of lipid peroxidation cytochrome P-450 2B4 exhibits a low oxidase activity in reconstituted membrane vesicles but still causes lipid peroxidation [6]. The ethanol-inducible isozyme cytochrome P-450 2E1, on the other hand, exhibits a high oxidase activity [18]. For this isozyme which is inducible by oxygen (95% oxygen atmosphere) too, its amount, generation of superoxide anion radical, and production of hydrogen peroxide and lipid peroxidation are correlated. Therefore this isozyme contributes to the oxygen-mediated tissue toxicity not only by its intrinsic lipid peroxidation but, moreover, by its inducibility by oxygen.

References

1. Bernhardt R, Makower A, Jänig G-R, Ruckpaul K (1984) Selective chemical modification of a functionally linked lysine in cytochrome P-450 LM2. Biochim Biophys Acta 785:186–190
2. Blanck J, Rein H, Sommer M, Ristau O, Smettan G, Ruckpaul K (1983) Correlationship between spin equilibrium shift, reduction rate, and N-demethylation activity in liver microsomal cytochrome P-450 and a series of benzphetamine analogues as substrates. Biochem Pharmacol 32:1683–1688
3. Blanck J, Smettan G, Ristau O, Ingelman-Sundberg, Ruckpaul, K (1984) Mechanisms of rate control of the NADPH-dependent cytochrome P-450 reduction by lipids in reconstituted phospholipid vesicles. Eur J Biochem 144:509–513
4. Blanck J, Ristau O, Zhukow AA, Archakov AI, Rein H, Ruckpaul K (1991) Cytochrome P-450 spin state and leakiness of the monooxygenase pathway. Xenobiotica 21:121–135
5. Brunström A, Ingelman-Sundberg M (1980) Benzo[a]pyrene metabolism by purified forms of rabbit liver microsomal cytochrome P-450, cytochrome b_5 and epoxide hydrase in reconstituted phospholipid vesicles. Biochem Biophys Res Commun 95:431–439
6. Ekström G, Ingelman-Sundberg M (1986) Mechanisms of lipid peroxidation dependent upon cytochrome P-450 LM2. J Biochem 158:195–201
7. Gonzales FJ (1990) Molecular genetics of the P-450 superfamily. Pharmacol Ther 45:1–38
8. Gorsky LD, Koop DR, Coon MJ (1984) On the stoichiometry of oxidase and monooxygenase reactions catalyzed by liver microsomal cytochrome P-450. J Biol Chem 259:6812–6817

9. Groot de H, Anundi I, Sies H (1991) Cytochrome P-450 and lipid peroxidation. In: Ruckpaul K, Rein H (eds) Frontiers in biotransformation, vol 5. Akademie, Berlin, pp 184–205
10. Rein H, Ristau O, Friedrich J, Jänig, G-R, Ruckpaul K (1977) Evidence of the existence of a high spin low spin equilibrium in liver microsomal cytochrome P-450. FEBS Lett 75:19–22
11. Ristau O, Rein H, Jänig, G-R, Ruckpaul K (1978) Quantitative analysis of the spin equilibrium of cytochrome P-450 LM2 fraction from rabbit liver microsomes. Biochim Biophys Acta 536:226–234
12. Ristau O, Wagnerova DM, Rein H, Ruckpaul K (1989) Demethylation of tertiary amines by a reconstituted cytochrome P-450 enzyme system: kinetics of oxygen consumption and hydrogen peroxide formation. J Inorg Biochem 37:111–118
13. Ruckpaul K, Rein H (eds) (1984) Cytochrome P-450, structural and functional relationships, biochemical and physicochemical aspects of mixed function oxidases. Akademie, Berlin
14. Ruckpaul K, Rein H, Blanck J (1989) Regulation mechanisms of the activity of the hepatic endoplasmic cytochrome P-450. In: Ruckpaul K, Rein H (eds) Frontiers in biotransformation, vol 1. Akademie, Berlin, pp 1–65
15. Ryan DE, Levin W (1990) Purification and characterization of hepatic microsomal cytochrome P-450. Pharmacol Ther 45:153–239
16. Schwarze W, Blanck J, Ristau O, Jänig G-R, Pommerening K, Rein H, Ruckpaul K (1985) Spin state control of cytochrome P-450 reduction and catalytic activity in a reconstituted cytochrome P-450 LM2 system as induced by a series of benzphetamine analogues. Chem Biol Interact 54:127–141
17. Thorgeirsson SS, McManus ME, Glowinski IB (1984) Metabolic processing of aromatic amides. In: Mitchell JR, Horning MG (eds) Drug metabolism and drug toxicity. Raven, New York, pp 183–197
18. Tindberg N, Ingelman-Sundberg M (1989) Cytochrome P-450 and oxygen toxicity. Oxygen-dependent induction of ethanol inducible cytochrome P-450 (II E1) in rat liver and lung. Biochemistry 28:4499–4504

Lipid Peroxidation and Hepatocyte Damage in the Animal Model and in Human Patients

G. Poli, E. Chiarpotto, F. Biasi, E. Albano, O. Danni, and M.U. Dianzani

Membrane lipid peroxidation is an event whose both initiation and propagation are based on free radical reactions. Within the cell there are several sources of free radicals able to act as initiating species, i.e., endoplasmic reticulum, mitochondrial and plasma membranes, soluble enzymes [9]. Of course, the basal production of these often reactive intermediates can be enhanced by external factors like variation in oxygen tension, high-energy radiation, toxic chemicals, carcinogens, and air pollutants.

Despite the fact that various macromolecules other than polyunsaturated fatty acids (PUFAs) can be actual targets for free radical reactions [9], lipid peroxidation appears of at least theoretical primary importance because of the biological role played by PUFAs in the membranes, the eventual target of all the mechanisms of cell injury.

Lipid Peroxidation as a Cause of Cell Damage

Starting with the animal model, to support a causative role of lipid peroxidation in the pathogenesis of cell damage, three conditions must be fulfilled: (1) actual evidence of increased lipid oxidation due to a defined compound; (2) its early appearance with respect to cell damage, and (3) protection against cell damage by antioxidants. As regards the third statement we partially disagree with those authors who tend not to consider lipid peroxidation as a causative mechanism of tissue injury when the latter is not totally prevented by means of experimental treatment with antioxidants. All these considerations have so far been made about studies performed in the animal model and more precisely in isolated cells (rat hepatocytes) whose maximum life in suspension (4–5 h) is in any case the final limit of the system. A significant though incomplete protection of cell death by antioxidants at the end of the experimental time cannot be considered as a mere delaying effect, but it is an actual proof in favor of a role of lipid peroxidation reactions in bringing about the injury. An up to date list of the literature reports concerning a possible role of membrane lipid peroxidation in the genesis of hepatocyte irreversible damage is given in Table 1. The present scheme derives from a previous one [12] with several additions and modifications. It is generally accepted that the primary mechanisms by

Table 1. Involvement of lipid peroxidation in the events leading to hepatocyte death (tentative update to 1990)

Generally accepted	Still discussed	Mainly excluded
Carbon tetrachloride[a]	t-Butylhydroperoxide	Menadione
Trichlorobromomethane	Sodium vanadate	Paraquat
1,2-dibromoethane		Diquat
Chloroform		Benzyl viologen
Paracetamol		Diethylmaleate
Iron		Iodoacetamide
Adriamycin		Cd, Hg, Cu
Ethoxycumarine		H_2O_2 (second phase)
Allyl alcohol[a]		
H_2O_2 (first phase)		
Bromobenzene[a]		
Halothane[a]		

[a] In vivo evidence is also available for these compounds.

which an excess of free radical reactions exert cell damage can be different [11]. While a rich group of toxic substances, now also including bromobenzene, have been demonstrated in the isolated rat hepatocyte model to induce cell damage mainly through stimulation of lipid peroxidation, several compounds have by contrast not shown a primary involvement of such a mechanism in their free radical mediated cytotoxicity.

The most relevant difference between the two lists of pro-oxidant compounds is the present availability of corresponding in vivo proofs only for those drugs which demonstrate their toxicity through lipid peroxidation. In fact, very recent studies in the whole rat or mouse gave further support to the implication of the latter mechanism in the hepatotoxicity of carbon tetrachloride [5], allyl alcohol [10], and bromobenzene [10].

All these data confirm that: (1) tissue damage due to free radicals, usually referred to as oxidative stress, does not necessarily imply lipid peroxidation since lipids are just one of the possible targets of these reactive chemical species and (2), at least in experimental toxic liver injury, a primary involvement of lipid peroxidation frequently appears highly likely.

Lipid Peroxidation and Derangement of Calcium Homeostasis

Over recent years the role of these two biochemical mechanisms in the genesis of irreversible hepatocyte damage has been extensively investigated and debated [2, 3]. Scientists often found themselves supporting one rather than the other process, which could be explained by the unique employment of models of acute toxicity. Most likely, in chronic liver intoxication the pathogenesis of the major events appears more complicated and the various mechanisms of damage more closely interrelated.

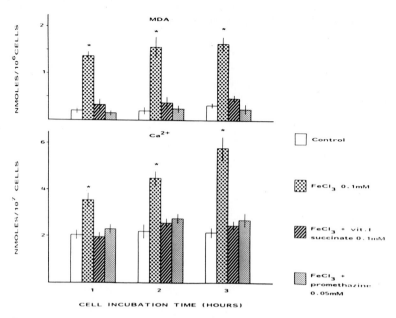

Fig. 1. Lipid peroxidation (*upper panel*) and calcium accumulation (*lower panel*) in isolated hepatocytes incubated at 37°C up to 3 h with or without 0.1 mM $FeCl_3$ in the presence or absence of 0.1 mM vitamin E succinate or 0.05 mM promethazine. Values are means ±SD of three experiments in duplicate. * Significantly different from the control ($p < 0.01$)

Very recently, Recknagel et al. [13] reassessed the possibility of an actual relationship between lipid peroxidation and intracellular rise of cytosolic calcium also in acute hepatocyte injury. These authors rather consider the derangement of calcium homeostasis as a second messenger of liver injury, which is triggered, at least in the case of carbon tetrachloride (CCl_4) intoxication, by lipid peroxidation and/or covalent binding of CCl_4 radical metabolites [14, 8]. In relation to this problem, it seems worth mentioning that, in the isolated hepatocyte system, cell enrichment in α-tocopherol did protect against cell death due to CCl_4, but failed to prevent the haloalkane-induced derangement of calcium homeostasis [1], while, on the contrary, the same antioxidant totally inhibited the rise of cytosolic calcium as a result of the acute iron overload [4] (Fig. 1).

In the light of our present knowledge we can draw the following conclusions: (1) an intracellular calcium rise could be due to the oxidation of protein thiols and of important cofactors [11], but it could also be the consequence of membrane lipid peroxidation; (2) an increase in intracellular calcium is not necessarily implied as the only second messenger of damage; (3) the actual relationship between this and other pathological mechanisms should be better characterized in vivo, especially in chronic models of intoxication.

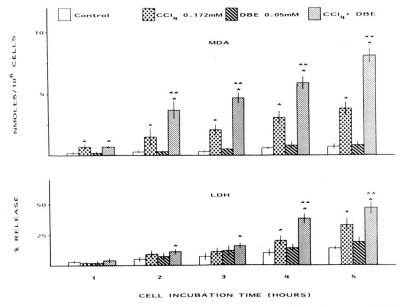

Fig. 2. Malonaldehyde (*MDA*) production (*upper panel*) and lactate dehydrogenase (*LDH*) release (*lower panel*) in isolated hepatocytes incubated at 37°C up to 5 h in the presence or absence of CCl₄ 0.172 mM, DBE 0.05 mM, or CCl₄ plus DBE. Values are means ±SD of three experiments in duplicate. * Significantly different from the control ($p < 0.05$); ** significantly different from CCl₄ ($p < 0.05$)

Interaction Among Pro-oxidant Agents

While much attention has been recently paid to the interaction between natural and/or synthetic antioxidants, little is still known about possible synergisms between pro-oxidant compounds, which are often present in the environment in mixtures rather than singly.

In relation to this subject, the stimulation of hepatocyte lipid peroxidation induced by CCl₄ acute poisoning in vitro has been shown to be potentiated by the simultaneous addition of per se non-pro-oxidant amounts of 1,2-dibromoethane (DBE). As reported in Fig. 2, malonaldehyde (MDA) production, taken as the index of lipid peroxidation, was increased by CCl₄ as soon as 60 min after incubation with isolated rat hepatocytes; a significantly higher increase in MDA formation was monitored in the presence of the two drugs 120 min after poisoning. Moreover, the CCl₄-DBE combination led to a significant potentiation of the membrane-damaging effect exerted by CCl₄. In fact, the release of cytoplasmic enzymes like lactate dehydrogenase (LDH) was enhanced when DBE was combined with CCl₄ (Fig. 2). The very good chronological correlation between the onset of increased MDA production (within the first 2 h) and the significant appearance of cell LDH leakage (after 2–3 h) was noteworthy. A causative role of

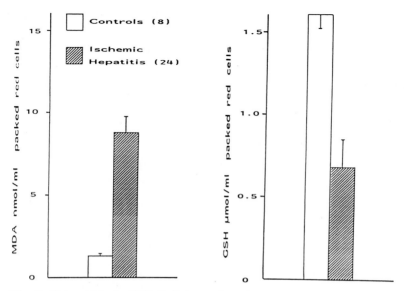

Fig. 3. Malonaldehyde (*MDA*) (*left panel*) and total glutathione (*GSH*) (*right panel*) levels in erythrocytes from normal subjects and patients with ischemic hepatitis. Number of cases is indicated in parentheses. Values are means ±SD. The difference between the two groups is significant at $p < 0.01$

lipid peroxidation in bringing about the potentiated effect of CCl_4 on cell irreversible damage was then confirmed by performing the described model of poisoning in vitamin E-enriched rat hepatocytes.

The supplementation of the antioxidant defenses allowed total prevention of the lipoperoxidative action of CCl_4 either alone or in combination with DBE [7]. Under such conditions, the irreversible cell damage due to CCl_4 was once again completely prevented, while that dependent upon the two drugs in combination was strongly even if not totally inhibited [7]. The predominant view at present on the mechanism(s) of the observed synergism supports a CCl_4-induced shift of DBE metabolism from its disposition through the formation of GSH adducts (via GSH transferase activity) to its cytochrome P-450 dependent activation.

Evidence of Increased Lipid Peroxidation in Human Liver Diseases

Two methods for in vivo detection of lipid peroxidation reactions are at present in our opinion precise and sufficiently reliable, i.e., erythrocyte MDA levels and fluorescent adducts between plasma proteins or lipids and aldehydic products of lipid peroxidation. As regards the latter measurement, an increased level of protein aldehyde and lipid aldehyde adducts has been shown in the plasma of human alcoholics [6]. The carbonyl compounds probably involved are MDA and 4-hydroxynonenal, among the major aldehydic products of membrane lipid peroxidation, which mainly react with nucleophilic groups of albumin and phospholipids. Most recently, in our laboratory, ischemic hepatitis has been considered as a human model for the verification of a causative role of lipid peroxidation in hepatocyte death. In fact, in such disease, which can be included in the ischemia-reperfusion pathology, increased levels of lipid peroxidation in terms of a significant rise as to normal subjects of both red cell MDA production (Fig. 3) and plasma fluorescent adducts (data not reported) have been observed. In addition, a big decrease of red cell total glutathione has been shown (Fig. 3), which supports the derangement of the antioxidant defenses. These findings parallel the characteristic very high levels of serum transaminases and lactate dehydrogenase. A conclusive demonstration of a relationship between the oxidative stress and necrosis described in ischemic hepatitis will probably be achieved by the clinical trials with antioxidants which are now in progress.

Acknowledgements. The authors wish to thank the Ministero dell' Università e della Ricerca Scientifica e Tecnologica and the Consiglio Nazionale delle Ricerche, Rome, for supporting this research.

References

1. Albano E, Bellomo G, Carini R, Biasi F, Poli G, Dianzani MU (1985) Mechanisms responsible for carbon tetrachloride-induced perturbation of mitochondrial calcium homeostasis. FEBS Lett 192:184–188
2. Albano E, Carini R, Parola M, Bellomo G, Poli G, Dianzani MU (1989a) Increase in cytosolic free calcium and its role in the pathogenesis of hepatocyte injury induced by carbon tetrachloride. In: Poli G, Cheeseman KH, Dianzani MU, Slater TF (eds) Free radicals in the pathogenesis of liver injury. Pergamon, Oxford, pp 45–54 (Advances in the biosciences, vol 76)
3. Albano E, Carini R, Parola M, Bellomo G, Goria-Gatti L, Poli G, Dianzani MU (1989b) Effects of carbon tetrachloride on calcium homeostasis. A critical reconsideration. Biochem Pharmacol 38:2719–2725
4. Albano E, Bellomo G, Parola M, Carini R, Dianzani MU (1991) Stimulation of lipid peroxidation increases the intracellular calcium content of isolated hepatocytes. Biochim Biophys Acta 1091:310–316
5. Biasi F, Albano E, Chiarpotto E, Corongiu F, Pronzato MA, Marinari UM, Parola M, Dianzani MU, Poli G (1991) In vivo and in vitro evidence concerning the role of lipid peroxidation in the mechanism of hepatocyte death due to carbon tetrachloride. Cell Biochem Funct 9:111–118
6. Carini R, Mazzanti R, Biasi F, Chiarpotto E, Marmo G, Moscrella S, Gentilini P, Dianzani MU, Poli G (1988) Fluorescent aldehyde-protein adducts in the blood serum of healthy alcoholics. In: Nordmann R, Ribiere C, Rouach H (eds) Alcohol toxicity and free radical mechanisms. Pergamon, Oxford, pp 61–64 (Advances in the biosciences, vol 71)
7. Danni O, Chiarpotto E, Aragno M, Biasi F, Comoglio A, Belliardo F, Dianzani MU, Poli G (1991) Lipid peroxidation and irreversible cell damage: synergism between

carbon tetrachloride and 1,2-dibromo-ethane in isolated rat hepatocytes. Toxicol Appl Pharmacol is 110:216–222

8. Dianzani MU, Poli G (1985) Lipid peroxidation and haloalkylation in CCl₄-induced liver injury. In: Poli G, Cheeseman KH, Dianzani MU, Slater TF (eds) Free radicals in liver injury. IRL, Oxford, pp 149–158

9. Freeman BA, Crapo JD (1982) Biology of disease. Free radicals and tissue injury. Lab Invest 47:412–426

10. Maellaro E, Casini AF, Del Bello B, Comporti M (1990) Lipid peroxidation and antioxidant systems in the liver injury produced by glutathione depleting agents. Biochem Pharmacol 39:1513–1521

11. Orrenius S (1988) Oxidative stress and hepatocyte damage. In: Poli G, Cheeseman KH, Dianzani MU, Slater TF (eds) Free radicals in the pathogenesis of liver injury. Pergamon, Oxford, pp 3–4 (Advances in the biosciences, vol 76)

12. Poli G, Albano E, Dianzani MU (1987) The role of lipid peroxidation in liver damage. Chem Phys Lipids 45:117–142

13. Rechnagel RO, Glende EA Jr, Dolak JA, Waller RL (1989) Mechanisms of carbon tetrachloride toxicity. Pharmacol Ther 43:139–154

14. Slater TF (1984) Free radical mechanisms in tissue injury. Biochem J 222:1–15

Calcium, Phospholipase A_2, and Eicosanoids in Toxigenic Liver Cell Injury

R.O. RECKNAGEL† and E.A. GLENDE, JR.

It is well known that the liver can be damaged by a wide variety of primary hepatotoxins. These include aliphatic and aromatic halogenated hydrocarbons; alcohols; quinones; bipyridyls; sulfur compounds; complex organic compounds of biological origin, e.g., phalloidin and aflatoxin; heavy metals, including iron; rare earths; elemental phosphorus; and many others. During the last 25 years a large and impressive body of knowledge has accumulated regarding the initial biochemical transformations undergone by xenobiotics in the liver. See [24] and [47] for reviews. As is well known the cytochrome P-450 mixed function oxidase system of the endoplasmic reticulum (ER) plays a prominent role in the initial reactions. As a consequence of the initial reactions, classical end-stage pathological phenomena ultimately appear: fatty degeneration, degranulation, necrosis, cancer, etc. [76]. The initial bioactivation reactions do not "cause" the appearance of end-stage pathological phenomena directly. Rather, the end-stage phenomena result from poorly understood secondary cascades of biochemical and cell physiological mechanisms set into motion by the initial events. The intermediate secondary mechanisms have proven difficult to unravel. For example, for any of the necrotizing hepatotoxins a complete description is not yet available for the events leading from initial bioactivation of the toxigenic agent to demise of the cell. To provide such a complete description turns out to be exceedingly difficult. Not only is any one chain of secondary events complex in and of itself, but there are more than one of such chains, and, no doubt, they act and react on each other. It seems reasonable to assume that an end result as complex as cellular necrosis could come about from a variety of pathological processes. Presumably, any one of such processes, if left unchecked, could result in death of the cell.

Free Radical Involvement and Lipid Peroxidation

In a wide variety of cases, toxigenic cell injury involves radical mechanisms [16, 24, 47]. It has been pointed out that, as details of the intermediate reactions are clarified, numerous related processes are opened up to closer scrutiny and grounds are thus created for possible rational therapy; see Preface to [16]. In this connection lipid peroxidation has attracted a great

Table 1. Xenobiotics causing hepatic lipid peroxidation (compiled January 1989)

Halogenated hydrocarbons—
CCl_4, $CBrCl_3$, $CHCl_3$, halothane, 1,2-dibromoethane, 1,1,2,2-tetrachloroethane,
  bromobenzene, iodobenzene
Alcohols—
Ethanol, propanol, isopropanol, allyl alcohol
Hydroperoxides—
Cumene hydroperoxide, t-butyl hydroperoxide
Redox cycling compounds
Paraquat, doxorubicin hydrochloride (Adriamycin)
Other compounds—
Acetaldehyde, acrylonitrile, cadmium, cocaine, endotoxin, ethoxycoumarin, paracetamol
  (acetaminophen), 2-(ethylmercurimercapto)-benzoic acid, p-phenylenediamine,
  iodoacetamide, N-chloro-p-toluenesulfonamide (chloramine-T), diethylmaleate,
  sodium vanadate

deal of attention. A list of compounds known to cause peroxidative decomposition of liver lipids is given in Table 1; the list is not intended to be exhaustive.

References to the scientific literature from which the list in Table 1 was compiled may be found in Recknagel et al. [47]. The wide range of inorganic and organic functional groups is noteworthy and implies the existence of a variety of different mechanisms that can lead to the lipid peroxidation, as indeed is known to be the case [24, 47]. Thus, lipid peroxidation can be initiated by carbon-centered or oxygen-centered free radical derivatives of xenobiotics via attack on bis-allelic hydrogens of polyenoic fatty acid side chains of phospholipids [16, 24, 47]. Through mechanisms not fully understood, lipid peroxidation can eventuate from partially reduced species of molecular oxygen arising during redox cycling of certain xenobiotics. It can also occur as a consequence of toxigenic compromising of mechanisms involved in defense against lipid peroxidation, e.g., by lowering of stores of reduced glutathione (GSH), and perhaps also by accelerated decomposition of low steady-state levels of lipid hydroperoxides, as may be taking place in the liver injury of hemochromatosis and experimental iron overload [5].

For any particular hepatotoxin, once it can be established that lipid peroxidation is involved, a range of new opportunities presents itself for further analysis of the overall problem. Properly speaking, lipid peroxidation should be regarded as only one link (albeit a most critical and important link) in the overall chain of causality leading from bioactivation to end-stage pathology. Viewed in this way, the questions that arise can be seen to fall into a number of distinct categories. On one hand are questions that relate to the chemistry of the initiation and propagation of the lipid peroxidation as caused by the toxigenic agent or its metabolites. Closely related is the question of the intracellular locus of the lipid peroxidation: what subcellular structures are involved? Questions such as these, important and interesting as they may be, nevertheless look back on that segment of the overall chain of causality linking bioactivation to emergence of the lipid peroxidation. It appears to us as self-evident that experimental preoccupation with this segment of the chain, however fascinating such preoccupation may be, cannot in and of itself lead to an uncovering of the processes ultimately causing emergence of the end-stage pathological phenomena characteristic of the particular xenobiotic under investigation. Clearly, clarification of the latter requires study of the consequences of the lipid peroxidation.

Carbon Tetrachloride

Because of the wealth of detailed knowledge that was available in the mid-1970s regarding the events of CCl$_4$ bioactivation, the problem of secondary pathological mechanisms for this classical hepatotoxin had already emerged clearly by that time [45]. It is now well established that lipid peroxidation follows immediately after generation of the trichloromethyl radical ($\cdot$CCl$_3$) by reductive dehalogenation of CCl$_4$, catalyzed by a specific isozyme of cytochrome P-450 in the ER, and by the electron transport chain in the mitochondria [67]; see [46] for a recent critical review. Clearly, these initial events do not, in and of themselves, cause the classical end-stage sequelae of CCl$_4$ hepatotoxicity, viz., the triglyceride accumulation, poly-ribosomal disaggregation, depression of protein synthesis, cell membrane breakdown, and eventual cell death. Secondary chains of events lead to these consequences.

In our opinion it is extremely unlikely that the highly reactive and short-lived $\cdot$CCl$_3$ and trichloromethyl peroxyl ($\cdot$OOCCl$_3$) radicals arising from CCl$_4$ metabolism could migrate any significant distances from their intra-cellular sites of origin to produce pathological changes in other parts of the cell. Other mechanisms must be at work linking CCl$_4$ metabolism to CCl$_4$-dependent liver cell injury. Until recently the main lines of work in this area have been dominated by two ideas. One of these centers on a possible role for 4-hydroxyalkenals and other products of membrane lipid peroxidation. The 4-hydroxyalkenals form Michael addition compounds by conjugation with SH groups of proteins and other compounds. They probably contribute significantly to the breakdown of the ER in CCl$_4$ and bromobenzene toxicity [8, 9]. These compounds, however, are rapidly detoxified in the cell cyto-plasm by conjugation with GSH and by reduction to corresponding alcohols. For these reasons it has been questioned whether the 4-hydroxyalkenals could play a significant role linking CCl$_4$ bioactivation to CCl$_4$ end-stage pathology involving parts of the cell other than the ER. See [5, 44, 46, 47] for discussion of this subject.

In a parallel development, attention has been directed to the idea that disturbed Ca^{2+} homeostasis might be involved. According to this idea

[37, 43], as a consequence of the early attack by CCl_4 on the ER, and the associated loss of the Ca^{2+}-sequestering capacity of this intracellular organelle [33, 34, 37], the concentration of Ca^{2+} ions in the cytoplasm, $[Ca^{2+}]_i$, would rise. It was postulated [43] that as a consequence of the rise in $[Ca^{2+}]_i$ otherwise latent hydrolytic enzymes, in particular phospholipase A_2 (PLA_2), would be activated, and that such activation, and possibly also activation of latent cytosolic proteases, would be crucial in the chain of causality linking bioactivation of CCl_4 to death of the cell. This so-called "Ca^{2+} hypothesis" for CCl_4-dependent liver cell injury has been the focus of a considerable amount of work, some aspects of which are reviewed in what immediately follows.

Effect of CCl_4 on Free Calcium of Liver Cell Cytosol

Whether "toxicologically meaningful" concentrations of CCl_4 cause a rise in $[Ca^{2+}]_i$ in isolated hepatocytes is an important question for the study of halogenated hydrocarbon hepatotoxicity. This subject has recently been reviewed [46]. Briefly, earlier work [31] showed that when isolated hepatocytes in culture were treated with CCl_4 there was a dramatic and rapid conversion of glycogen phosphorylase b to the active a form, which implies a rise in $[Ca^{2+}]_i$. It was subsequently reported [32] for experiments on isolated hepatocytes loaded with quin-2 that $[Ca^{2+}]_i$ rose threefold within 20 s of addition of CCl_4. These experiments appeared to lend strong support to the idea that disturbed Ca^{2+} homeostasis may play a role in CCl_4-dependent liver cell injury. Their relevance, however, was questioned [13, 46] on the grounds that the estimated concentrations of CCl_4 used in these experiments were in the range of 2–3 mM. Such high concentrations are probably not "toxicologically meaningful" in the sense that effective concentrations of CCl_4 in vivo in the rat probably never reach levels of 2–3 mM. For example, after 2.5 ml CCl_4/kg body weight (per os), which is a large dose, the concentration of CCl_4 in portal blood of the rat did not rise above 0.6 mM (Glende E.A. Jr, unpublished). In experiments based on isolated hepatocytes loaded with fura-2, it was shown conclusively that high concentrations of CCl_4 exert a rapid direct permeabilizing effect on the hepatocyte plasma membrane [13]. Also, see ref. [4]. Such rapid permeabilizing effects are not relevant to pathological consequences of CCl_4 that depend on reductive dehalogenation of the CCl_4 molecule.

The fluorescence method that permits a judgement to be made as to whether a given concentration of CCl_4 is or is not causing direct permeabilizing effects emerges from the basic procedures routinely followed when fura-2 fluorescence is being used to determine $[Ca^{2+}]_i$. For experiments with fura-2-loaded cells, addition of Mn^{2+} ions is one of the crucial steps in obtaining the fluorescence readings necessary for calculation of $[Ca^{2+}]_i$; see [13] for details. If conditions prior to addition of Mn^{2+} ions have permeabilized the hepatocyte plasma membrane ·(e.g., high concentrations of CCl_4), on addition of Mn^{2+} ions there will be a large drop in the fluorescence signal as Mn^{2+} ions quench all fura-2 fluorescence, extracellular as well as intracellular, and there will be no further drop in the fluorescence signal on addition of digitonin; see Fig. 2 of ref. [13]. On the other hand, if prior conditions have not caused any permeabilization of the plasma membrane, on addition of Mn^{2+} ions there will be a modest drop in the fluorescence signal as Mn^{2+} ions quench only extracellular fura-2 fluorescence, and there will be a further large drop in fluorescence on addition of digitonin, which permeabilizes the cells and renders intracellular fura-2 available for Mn^{2+} ion quenching; see Fig. 1 of ref. [13]. With these methods, it was reported that theoretical "input" concentrations of $3\,mM$ and $5\,mM$ CCl_4 directly permeabilized the plasma membrane. A theoretical "input" concentration of $2\,mM$ CCl_4 produced marginal direct permeabilization, and "input" concentrations of $1\,mM$ and $0.5\,mM$ did not produce direct permeabilizing effects.

These experiments [13] were carried out in a 3-ml spectrophotofluorometer cuvette fitted with a plastic cap that was not gastight. Routinely, the cuvette contained $2\,ml$ fura-2-loaded hepatocytes at 37.5°C. Suspension of the hepatocytes was maintained by stirring with a magnetic stirring bar. CCl_4 was always added as a 5-μl aliquot of an appropriate solution of CCl_4 in dimethylsulfoxide (DMSO). The final concentration of CCl_4 was calculated on the assumption that all added CCl_4 remained in solution in the $2\,ml$ aqueous suspension of hepatocytes. Given the highly volatile nature of CCl_4 and the fact that the cuvette system was not gastight and had an air headspace volume of approximately $1\,ml$, it was recognized that actual concentrations of CCl_4 may have been lower than the calculated concentrations. This expectation has been born out in recent work in this laboratory. Use was made of $[^{14}C]$-CCl_4 to track the concentration of CCl_4 in a cuvette system patterned exactly after that of Dolak et al. [13]. For a theoretical "input" concentration of $5\,mM$ CCl_4 (i.e., the calculated concentration if all added CCl_4 remained dissolved in the $2\,ml$ aqueous phase) the actual concentration of CCl_4 in the aqueous phase reached a maximum of just over $2\,mM$ after 3 min of stirring and steadily declined thereafter, to $0.35\,mM$ by 30 min (Fig. 1). For a theoretical "input" concentration of $2\,mM$, the highest observed concentration of CCl_4 was $1.4\,mM$, which was reached in about 1 min, with a steady decline thereafter (data not shown).

In the experiments of Dolak et al. [13], theoretical "input" concentrations of 0.5 and $1.0\,mM$ CCl_4 did not produce direct permeabilizing effects, but they did cause a modest sustained rise of $[Ca^{2+}]_i$. The question remained as to what were the actual concentrations of CCl_4 that produced the rise in $[Ca^{2+}]_i$. For a theoretical "input" concentration of $1.0\,mM$, the $[^{14}C]CCl_4$ data showed that the concentration of CCl_4 in the aqueous phase was $1.09\,mM$ at 0.3 min of stirring with steady decline to $0.37\,min$ at 20 min (Fig. 2).

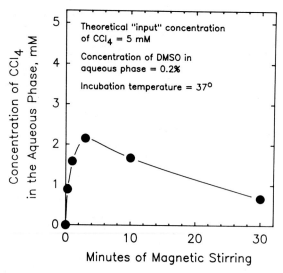

Fig. 1. Change in concentration of CCl₄ in a stirred aqueous phase. Calculated "input" concentration of CCl₄ was $5 \, mM$. *Conditions*: $5 \, \mu l$ of a $2.0 \, M$ solution of $[^{14}C]CCl_4$ in DMSO was added to $2.0 \, ml \, H_2O$ at $37.5°C$ in a spectrophotofluorometer cuvette. A plastic-coated magnetic stirring bar was added, the cuvette was capped, and the mixture was stirred by a remote magnet. At the times indicated, 0.5-μl aliquots of the mixture were removed for determination of radioactivity

For a theoretical "input" concentration of $0.5 \, mM$, the actual concentration was $0.5 \, mM$ at 0.3 min of stirring with a steady decline to $0.06 \, mM$ at 30 min (data not shown). Presence of hepatocytes had little or no effect on either the concentration of CCl₄ reached in the aqueous phase or on the rate of loss of CCl₄ wih stirring.

Our $[^{14}C]CCl_4$ data (Fig. 2) in conjunction with the data reported in [13] permit the conclusion that concentrations of CCl₄ up to $1 \, mM$ do not cause direct permeabilization of the hepatocyte plasma membrane. A CCl₄ concentration of $1.4 \, mM$ is marginal with respect to direct permeabilizing effects, whereas, on the basis of indirect criteria (see Fig. 3, below, et seq.), $1.2 \, mM$ was found to not produce direct permeabilizing effects. From the point of view of the "Ca^{2+} hypothesis" for CCl₄ liver cell injury it is significant that $1.0 \, mM$ and $0.5 \, mM$ CCl₄ caused a modest but sustained rise in $[Ca^{2+}]_i$ [13], but whether the modest rise in $[Ca^{2+}]_i$ plays a significant role in the chain of causality linking initial metabolic activation of CCl₄ to end stage pathological phenomena has not been demonstrated unequivocally. See Albano et al. [1] for a recent study of effects of CCl₄ on $[Ca^{2+}]_i$ of isolated hepatocytes in which it was reported that concentrations of CCl₄ of 0.35 and $0.52 \, mM$ caused a rise in $[Ca^{2+}]_i$. However, a low concentration of $0.172 \, mM$ did cause significant cell damage during 4 h of incubation, but did not cause elevation of $[Ca^{2+}]_i$ during the first 30 min of incubation, as

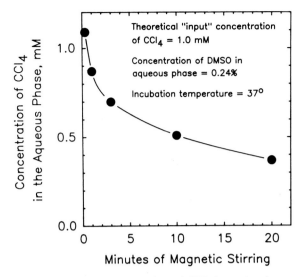

Fig. 2. Change in concentration of CCl$_4$ in a stirred aqueous phase. Calculated "input" concentration of CCl$_4$ was 1 mM. *Conditions*: Same as for Fig. 1, except 5 μl of a 0.4 M solution of [^{14}C]CCl$_4$ in DMSO was added to the 2.0 ml H$_2$O

determined by quin-2 fluorescence and activation of glycogen phosphorylase. A rise in [Ca^{2+}]$_i$ after 30 min of incubation with 0.172 mM CCl$_4$ was ascribed to entrance of external Ca^{2+} into the hepatocytes. It is evident that further work will be needed to settle unresolved questions as to whether toxicological effects of nonpermeabilizing concentrations of CCl$_4$ on isolated hepatocytes are dependent on disturbance of hepatocyte Ca^{2+} homeostasis.

Activation of Phospholipase A$_2$

One possible concequence of a rise in [Ca^{2+}]$_i$ is the activation of PLA$_2$. There is no doubt [18] that PLA$_2$ is activated during the course of CCl$_4$-dependent hepatocyte injury. Phospholipase A$_2$ activation can be tracked by appearance of ^{3}H label in the free fatty acid fraction of hepatocytes prelabeled wih [^{3}H]arachidonic acid [6, 18] or by appearance of ^{14}C-labeled lysophosphatidylethanolamine in hepatocytes prelabeled with [^{14}C]ethanolamine [18]. A typical result is shown in Fig. 3 (Glende E.A. Jr, unpublished).

Note that as early as 20 min after addition of the cells to the medium containing CCl$_4$ the content of [^{14}C]lysophosphatidylethanolamine in the total lipid extract of the cells has increased markedly, indicating that activation of PLA$_2$ has occurred. After about another 20 min of incubation, an

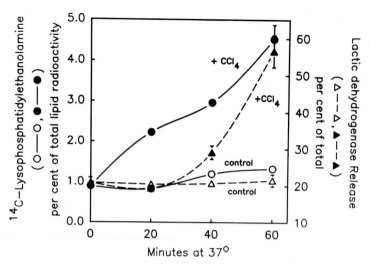

Fig. 3. Activation of phospholipase A_2 by CCl_4 in isolated hepatocytes. (Glende EA Jr, unpublished). *Conditions*: Hepatocytes, previously incubated with [1,2-^{14}C]ethanolamine, were incubated at 37°C in Waymouth's MB 752/1 medium (Gibco Laboratories) buffered at pH 7.4 with 25 mM HEPES in a stoppered 50-ml plastic flask. The incubation medium was equilibrated with CCl_4 (25 µl CCl_4 added to a 6 × 50-mm center tube) for 60 min at 37°C prior to addition of hepatocytes to a final suspension concentration of 2% (w/v) in 3.0 ml volume. The equilibrated aqueous concentration CCl_4 in this system as determined by gas chromatographic/electron capture analysis was 1.2 ± 0.1 mM. Aliquots of the incubation mixture were taken for determination of [^{14}C]lysophosphatidylethanolamine content by thin-layer chromatographic analysis (phospholipase A_2 activity) and extracellular lactic dehydrogenase activity (cell viability)

increase in medium lactic dehydrogenase can be detected, indicating that by this time some cell death has occurred.

The judgement that the concentration of 1.2 mM CCl_4 used (Fig. 3) was most probably not causing any direct membrane permeabilization is based on two points. First, hepatocyte injury (LDH release) was not evident until about 40 min of incubation. Cell injury effects from solvent action of CCl_4 are much more rapid, occurring in seconds [32] to a few minutes [4]. Second, for the experiment reported in Fig. 3 the medium concentration of Ca^{2+} was $2.5 \times 10^{-3} M$, but reduction of medium Ca^{2+} concentration to $1.0 \times 10^{-7} M$ did not alter the CCl_4-dependent activation of PLA_2 or the extent of cell injury (Glende E.A. Jr, unpublished). Thus, entrance of medium Ca^{2+} into the hepatocytes, as would occur as a result of direct plasma membrane permeabilization, is evidently not the cause of the effects shown in Fig. 3. Furthermore, PLA_2 activation in isolated hepatocytes was reported for concentrations of CCl_4 as low as 0.35 mM [18]. A similar time sequence, i.e., PLA_2 activation followed about 30 min later by the first evidences of cell death, was observed also for isolated hepatocytes intoxicated in vitro with $CBrCl_3$, $CHCl_3$, and vinylidene chloride (Glende E.A. Jr, unpublished).

Is Activation of Phospholipase A$_2$ Pathologically Significant?

There are at least two ways by which activation of PLA$_2$ could have pathological consequences. One of these is by generation of "toxic lipid mediators" in the arachidoniacid cascade (see below), and the other is via PLA$_2$-dependent catalytic hydrolysis of membrane phospholipids at rates greater than rates of membrane repair. Obviously, these two putative modes of pathological action of PLA$_2$ are not mutually exclusive.

Activation of PLA$_2$ plays a key physiological role in one aspect of GSH-dependent protection of membranes against lipoperoxidative damage. There is an absolute requirement that PLA$_2$ first cleave any peroxidized fatty aicd from a peroxidized membrane phospholipid before such peroxidized fatty acid can be reduced to its corresponding hydroxy fatty acid derivative by cytosolic GSH peroxidase. Repair of the membrane is effected by reacylation of the lysophospholipid with an appropriate long-chain fatty acid acyl CoA derivative [60, 73], and see [47] for a general discussion of mechanisms operative in protection of membranes against lipoperoxidative injury.

Apart from any physiologically meaningful role for PLA$_2$ activity, the fact that PLA$_2$ is activated by CCl$_4$ and other hepatotoxins raises the question as to whether toxigenic activation of this enzyme could have significant pathological consequences. It has of course been known for many years that PLA$_2$ activity figures prominently in the toxicity of bee venom [11] and certain snake venoms [23]. Poisoning by such venoms is a question of introduction into an organism of exogenous PLA$_2$ by insect sting or snake bite. Although toxigenic activation of otherwise latent endogenous PLA$_2$ is mechanistically a different problem, it seems not unlikely that such activation could have deleterious effects. For example, in cultured fibroblasts hydrolysis of as little as 15% of phospholipids was associated with death of all cells [61]. Alteration of hepatic microsomal lipids by the action of PLA$_2$ significantly depresses glucose-6-phosphatase activity and calcium-sequestering capability [7, 14]. Since alteration of membrane phospholipids by lipid peroxidation promotes an increase in PLA$_2$ activity [42, 73–75], it seems reasonable to suppose that during active lipid peroxidation hydrolysis of phospholipids at rates greater than rates of membrane repair would contribute to the membrane breakdown [15].

In a study of isolated rat hepatocytes exposed to CBrCl$_3$ (experiments with ADP-Fe^{2+} were also reported) it was shown that collapse of plasma membrane function was delayed until about 60 min after the time of peak lipid peroxidation, which occurred at about 30 min. At the time of collapse of plasma membrane function there were significant decreases of plasma membrane phosphatidylcholine and phosphatidylethanolamine with concomitant increases of their respective lysophosphatides [70–72]. Sudden loss of functional membrane properties correlated with a steep rise of plasma membrane lysophosphatides, and not with peak lipid peroxidation, which occurred earlier. The temporal correlation of increased membrane lysophosphatides with loss of membrane functional properties is encouraging for the view that activation of PLA_2 plays a role in $CBrCl_3$ toxicity.

One way of attempting to assess a possible role for PLA_2 activation in the mechanism of action of hepatotoxins is to attempt to block or at least to attenuate the toxicological effects by use of inhibitors of PLA_2. Such attempts have usually resorted to use of cationic amphiphilic substances such as chlorpromazine and quinacrine. Thus, it was reported [70, 71] for the experiments with $CBrCl_3$ and $ADP-Fe^{2+}$ that if PLA_2 of hepatocytes was inhibited by either of these substances the sudden collapse of plasma membrane function that occurs about 60 min after the time of peak lipid peroxidation was markedly attenuated, and there was marked preservation of cell integrity [72]. Although these results can be taken as support for a role for PLA_2 activation in the necrotizing effects of $CBrCl_3$, the conclusion is tempered by a number of considerations. For one thing, the K_i values for these substances (inhibition of PLA_2 of porcine pancreas) are high: 345 and 1213 µmol/l, respectively, for chlorpromazine and quinacrine. Hepatocytes tolerate these substances poorly, i.e., input concentrations necessary to achieve inhibition of hepatocyte PLA_2 are themselves significantly hepatotoxic. Furthermore, these PLA_2 inhibitors exert antioxidant action [72], which complicates interpretations when lipid peroxidation is involved in the mechanism of action.

The possibility that activation of PLA_2 plays a role in the necrotizing effects of $CBrCl_3$ was lent further strong support by experiments carried out with hepatocytes enriched with nonperoxidizable dipalmitoyl-lecithin (DPL). It was reported [72] that, whereas up to 20% of total phospholipids of control hepatocytes were converted to lysophospholipids by $CBrCl_3$ treatment, the conversion was only 8% for hepatocytes enriched with DPL, and the decrease in $CBrCl_3$-dependent generation of lysophosphatides due to DPL enrichment resulted in a marked attenuation of cell injury. Furthermore, for the cells enriched with DPL, there was no diminution of $CBrCl_3$-dependent lipid peroxidation. These results support the idea that activation of PLA_2 plays a role in the necrotizing action of $CBrCl_3$.

It is well known that there are close similarities in the mechanisms of action of $CBrCl_3$ and CCl_4 [29, 63]. By extension to CCl_4 of the results for $CBrCl_3$ [70–72] from the known deleterious effects on biomembranes of the uncontrolled action of PLA_2, and from the fact that PLA_2 is known to be activated when hepatocytes are exposed to CCl_4 [6, 18], it seems reasonable to conclude that PLA_2 activation plays a role in CCl_4-dependent liver cell injury. This formulation should not be taken to mean that activation of PLA_2 is a necessary requirement for expression of the necrotizing potential of CCl_4 or any other hepatotoxin known to activate PLA_2. Support for the latter idea has emerged from work in a number of laboratories. Chiarpotto et al. [6] showed that indomethacin and mepacrine (quinacrine), both of which are PLA_2 inhibitors, significantly inhibited the CCl_4 activation of PLA_2 in isolated hepatocytes, but neither agent had any effect in preventing

CCl$_4$-dependent cell death. Albano et al. [1] reported that a series of PLA$_2$ and protease inhibitors were ineffective in preventing LDH leakage and surface blebbing in rat hepatocytes treated in vitro with $0.172\,mM$ CCl$_4$. In a study of cystamine hepatotoxicity, Nicotera et al. [40] reported that the PLA$_2$ inhibitors chlorpromazine and dibucaine effectively inhibited PLA$_2$ activation, but had no protective action on cell viability. Significantly, the cytosolic protease inhibitors leupeptin and antipain did prevent cystamine-dependent cell death, indicating that for this hepatotoxin activation of cytosolic protease activity is probably critical for expression of lethality. Stacey and Klaassen [64] reported that hepatocyte necrosis due to diethylmaleate was not prevented by either chlorpromazine or quinacrine. Similarly, Tsokos-Kuhn [69], in a study of diquat hepatotoxicity, reported that whereas chlorpromazine effectively inhibited PLA$_2$ activation it did not exert complete protection against cell death. However, as emphasized above, cell death can no doubt ensue from a variety of causes. If cell death occurs despite effective inhibition of PLA$_2$, then causes other than PLA$_2$ activation must be involved. Nevertheless, it seems to us rather likely, given the destructive potentiality of unregulated PLA$_2$ activity, that for any hepatotoxin under investigation, if PLA$_2$ activation occurs, then such activation, if not inhibited, will probably contribute to the overall pathological consequences. What has emerged from recent work is that, at least for some hepatotoxins known to activate PLA$_2$, the PLA$_2$ activation does not appear to be an indispensable link in the chain of causality leading from the initial events of xenobiotic metabolism to ultimate cellular necrosis.

Role of Eicosanoids in Toxigenic Liver Cell Injury

A rapidly expanding aspect of the study of xenobiotic hepatotoxicity involves study of protective effects exerted by certain prostaglandins (PGs), and the generation via PLA$_2$ activation of eicosanoids, some of which exert direct and/or indirect toxic effects on hepatocytes. It is not possible here to offer a complete review of this complex subject; however, a few comments with guides to some of the pertinent literature may be in order.

Prostaglandin Cytoprotection

It was reported in the mid-1970s that PGF$_{2\alpha}$ and PGF$_{2\beta}$, which do not affect gastric acid secretion, nevertheless prevent indomethacin-induced gastric ulceration [49]. It was observed subsequently that other PGs exert antiulcer activity at doses far too low to significantly inhibit gastric acid secretion, and that certain PGs protect the stomach against necrotizing agents that do not act via stimulation of gastric acid secretion. The phenomenon was called "cytoprotection." The protection has a rapid onset and is exerted against a widely disparate array of gastric irritants. Thus, PGA$_2$,

PGE_2, $PGF_{2\alpha}$, $PGF_{2\beta}$, and PGI_2 prevented gastric ulceration caused by absolute ethanol, $0.6N$ HCl, $0.2N$ NaOH, 25% NaCl, acidified aspirin or taurocholate, and boiling water [50]. Interestingly, mild gastric irritation protected against subsequent impositon of a necrotizing irritation by a mechanism inhibitable by indomethacin, implying that the "adaptive cyto-protection" of mild irritation involves generation of protective PGs. By the early 1980s PG cytoprotection was extended to toxigenic liver cell injury, with the reports (based on work in vivo) that 16,16-dimethyl PGE_2 ($dmPGE_2$) exerted remarkable to more modest protection of liver cells against the hepatotoxic effects of CCl_4, galactosamine, acetaminophen, ethanol, α-naphtholisothiocyanate (ANIT), and aflatoxin [52–54, 56, 57, 65, 66]. It was soon shown that $dmPGE_2$ protection against acute CCl_4-dependent liver cell injury is also expressed in vitro [55, 58]. PG cytoprotec-tion is not limited to xenobiotic toxicity since it has been shown that prostacyclin (PGI_2) exerts marked protection against hypoxic cell damage [2].

The mechanism of the cytoprotection is unknown. The fact that the cytoprotective PGs exert their protective effects in such a wide variety of toxigenic and pathogenic situations, with widely differing underlying biochemical mechanisms of cell injury, implies intervention by the cyto-protective PGs on some late-occurring event or events in the cascade of intermediary mechanisms linking early events of toxigenic or pathogenic insult to ultimate irreversible cell injury. In a careful study [17] it was reported that $dmPGE_2$ exerted remarkable protection against the hepato-toxic effects of bromobenzene administration to mice: $dmPGE_2$ reduced bromobenzene-dependent serum alanine aminotransferase activity and hepatocellular necrosis. It attenuated bromobenzene-dependent depletion of GSH and disappearance of cytochrome P-450; and it significantly reduced bromobenzene-dependent mortality. However, $dmPGE_2$ administration did not modify bromobenzene clearance from plasma and liver, and it did not modify covalent binding of bromobenzene metabolites to hepatic proteins. These results again suggest an action of $dmPGE_2$ on some late-developing critically important consequence of the early metabolism of this hepatotoxic xenobiotic.

In the case of CCl_4 liver cell injury, it seems unlikely that $dmPGE_2$ protection depends significantly on early events in the cell-CCl_4 response mechanism since the effect of three subcutaneous doses of $dmPGE_2$ (two given the day before and a third given 30 min prior to CCl_4 administration to rats), although cytoprotective, had only modest effects on a number of indices of overall CCl_4 metabolism [51]. Cytoprotective effects of PGI_2 against CCl_4 liver cell damage also appear to involve a late-phase action. Thus, it was reported [12] that PGI_2 and two of its derivatives exerted considerably greater cytoprotection when they were given 24 h after rather than 1 h before administration of CCl_4 to rats, a result that clearly points to action of these cytoprotective PGs on a late phase of the injury. At the in vitro level, PGI$_2$ exerts protection when given 30 min after exposure of isolated hepatocytes to CCl$_4$ [20, 21, 36]. Since the key events of trichloromethyl radical generation, initiation of lipid peroxidation, and disturbance of Ca^{2+} homeostasis are all fully developed in vitro well before 30 min, protection by PGI$_2$ when added at 30 min implies an effect on a late-developing aspect of the CCl$_4$ injury. The fact that PGI$_2$ is not protective when given 3 h after exposure of isolated hepatocytes to CCl$_4$ [20, 21] is not too surprising, since, for hepatocytes exposed to CCl$_4$ in vitro, by 3 h the progression of the necrotizing cascade has most probably reached an irreversible stage. It is interesting to note [see 36] that, if isolated rat hepatocytes were exposed to a low concentration of ethanol (1.5 µg/ml) for 30 min prior to exposure to CCl$_4$, there was a significant attenuation of hepatocyte injury. Also, the low concentration of ethanol increased hepatocyte production of 6-keto-PGF$_{1\alpha}$. These results imply an "adaptive cytoprotection" by the low dose of ethanol, a conclusion strongly supported by the further observation that the ethanol "adaptive cytoprotection" was lost in the presence of indomethacin.

Thanks to an enormous amount of work by many investigators, study of basic biochemical mechanisms in hepatic metabolism of xenobiotics may be said to be relatively advanced [24, 25]. Further, study of secondary cascades of pathological mechanisms in toxigenic liver cell injury is also relatively advanced. Thus, the consequences of the metabolism of primary hepatotoxins on Ca^{2+} homeostasis [3, 46], on hepatic pools of GSH [48], on activation of PLA$_2$ [18] and cytosolic proteases [40], and on hepatocyte blebbing [19, 30] have been extensively studied. Since the action of the protective PGs does not seem to be exerted on the early events of xenobiotic metabolism, study of PG modulation of the secondary mechanisms set into motion by the early metabolic events will no doubt lead to new insights into the mechanisms ultimately responsible for cellullar necrosis. See below for a discussion of effects of thromboxane B$_2$ on hepatocyte bleb formation and activation of a nonlysosomal protease.

Active Role of Eicosanoids: Leukotrienes

In addition to prostaglandin protective effects, products of the arachidonic acid cascade can play a direct role in the overall response of the liver to toxigenic xenobiotics. For example, leukotrienes (products of the 5-lipoxygenase pathway) play a role in the overall response of the liver to experimental procedures that result in fulminant hepatitis; see [28] for a review. Liver injury in mice treated with lipopolysaccharide endotoxin (LPS) and either α-amanitin or galactosamine was markedly attenuated by diethylcarbazine, which inhibits LTA$_4$ snythesis; by FPL 55712, a receptor antagonist for cysteinyl leukotrienes; by a single pretreatment dose of BW 755c, an inhibitor of arachidonate lipoxygenase; and by dexamethazone, which operates against arachidonate liberation by promoting synthesis and release of lipocortin, an antiphospholipase protein. As pointed out [28], this inhibitor profile suggests an involvement of leukotrienes in endotoxin-generated liver injury. See [38] for a similar inhibitor profile for LPS and acute CCl_4 liver injury in mice. Leukotriene involvement has also been implicated in chronic CCl_4 liver cell injury in mice [62].

The mechanism of leukotriene-dependent liver cell injury is undoubtedly complex. It is well known that leukotrienes are predominantly produced by macrophages, monocytes, neutrophils, eosinophils, and mast cells; see Table 1 of [28]. Kupffer's cells isolated from normal rat liver release peptide leukotrienes in response to A23187 [59]. Also, see [10]. The involvement of leukotrienes in toxigenic liver injury is probably best envisaged from the point of view of the inflammatory reaction. In some way, as a response to primary hepatocyte injury, Kupffer's cells, and perhaps other cells in the liver, release leukotrienes. The chemotactic and chemokinetic action of LTB_4 results in leukocyte accumulation. The peptide leukotrienes LTC_4, LTD_4, and LTE_4 increase vascular permeability, allowing easier escape of phagocytic leukocytes into the interstitial compartment. Phagocytosis of cell debris from primary hepatocyte injury would lead to generation of superoxide anion radical in the well-known oxygen burst of leukocyte activation, with subsequent generation of H_2O_2, hydroxyl radical, hypochlorous acid, and other strong oxidants which can attack and injure nearby cells. Also, leukotrienes may play a more direct toxic role, since it has been shown that hepatocytes cultured at low oxygen concentrations are killed by nanomolar concentrations of LTC_4 [68].

Active Role of Eicosanoids: Prostanoids

It is clear from recent work that prostaglandins and thromboxanes are generated in the necrotizing response of hepatocytes to certain foreign substances. The characteristic pattern of work in these studies consists of imposition on liver cells of a toxigenic perturbation with subsequent observation of some end result, e.g., formation of plasma membrane blebs. Bleb formation precedes frank disintegration of the cell plasma membrane [19, 30]; it has been used as an early manifestation of cell injury [40, 41]. By a variety of methods and treatments an attempt is made to unravel the intermediate events linking the initial perturbations to the end result. These methods and treatments include attempting to detect an increase in cytosolic Ca^{2+}, determining whether PLA_2 and cytosolic protease activity have increased, use of inhibitors of PLA_2 and proteases, determining whether prostanoids have been generated, use of various inhibitors of prostaglandin and thromboxane biosynthesis, and testing of various prostanoids and other eicosanoids for protective effects and for toxicity. The simplified scheme shown in Fig. 4, based in part on a study of bleb formation caused by action of A23187 on isolated hepatocytes [26], is offered as a guide to the rationale for some characteristic interventions.

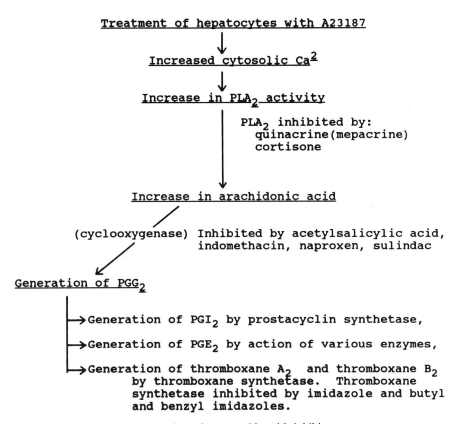

Treatment of hepatocytes with A23187

↓

Increased cytosolic Ca²

↓

Increase in PLA₂ activity

PLA₂ inhibited by:
quinacrine(mepacrine)
cortisone

↓

Increase in arachidonic acid

(cyclooxygenase) Inhibited by acetylsalicylic acid,
indomethacin, naproxen, sulindac

Generation of PGG₂

→ Generation of PGI₂ by prostacyclin synthetase,

→ Generation of PGE₂ by action of various enzymes,

→ Generation of thromboxane A₂ and thromboxane B₂
by thromboxane synthetase. Thromboxane
synthetase inhibited by imidazole and butyl
and benzyl imidazoles.

Fig. 4. Partial scheme of generation of prostanoids, with inhibitors

What has emerged from the recent work is the idea of the existence of a natural PGI_2-TXA_2,TXB_2 balance, in which the cytoprotective action of PGI_2 acts as a counterweight to the direct cytotoxic potentiality of the thromboxanes. In keeping with this idea, Guarner et al. [22; also, see 25] reported that an acetaminophen overdose in the mouse resulted in enhanced prostaglandin and thromboxane generation by liver homogenates. Prostacyclin treatment of the whole animal prevented acetaminophen-induced liver necrosis. This prostacyclin protective effect does not in and of itself indicate the existence of a PGI_2-TXA_2,TXB_2 balance. However, in liver homogenates, the selective thromboxane synthetase inhibitor OKY 1581 inhibited thromboxane production and increased production of prostacyclin, and, in the whole animal OKY 1581, given shortly after acetaminophen, prevented mortality and reduced liver necrosis. These results support the idea that acetaminophen toxicity upsets a natural PGI_2-TXA_2,TXB_2 balance in which the thromboxanes play a direct toxic role and PGI_2 is cytoprotective.

Horton and Wood [26] found that A23187-dependent bleb formation in isolated hepatocytes was markedly inhibited by quinacrine and cortisone, by the nonsteroidal antiinflammatory agents naproxen and sulindac, and by imidazole and butylamidazole, results which strongly implicate generation of TXA_2 and TXB_2 as toxic mediators in A23187 elicitation of plasma membrane blebs. Further, TXB_2 at $1 \times 10^{-5} M$ produced bleb formation in 76% of cells. PGE_2 was considerably less active and PGI_2 was without effect. Subsequently, it was shown that PGI_2 had a strong cytoprotective effect against TXB_2-induced hepatocyte bleb formation [27], as did $dmPGE_2$. These results strongly support the existence of a natural PGI_2-TXA_2,TXB_2 balance.

A particularly interesting observation of Horton and Wood [27] concerns an inhibitory action of leupeptin on TXB_2-induced plasma membrane bleb formation in isolated hepatocytes. Agents such as A23187 or cystamine, which cause an increase in $[Ca^{2+}]_i$ in isolated hepatocytes, also cause bleb formation and loss of cell viability. Bleb formation is believed to be intimately linked to a breakdown in the normal structure and function of the cytoskeleton. For example, bleb formation is stimulated by agents such as cytochalasin B or D and by phalloidin, which are believed to act directly on the cytoskeleton. The possible involvement of Ca^{2+} -activated neutral proteases in toxigenic bleb formation and hepatocyte death emerged primarily from work by Nicotera et al. [40, 41], who observed that agents such as A23187 and cystamine, which cause an increase in $[Ca^{2+}]_i$, also cause bleb formation and loss of cell viability, and that bleb formation and cell death are prevented when leupeptin or antipain, which are inhibitors of Ca^{2+}-activated neutral proteases, are present in the incubation medium. It was suggested that the cytoskeletal proteins might be the targets for the non-lysosomal proteases [41]. Horton and Wood [25, 26] had proposed a general mechanism for Ca^{2+}-dependent cell injury and death with a hypothetical progression from an increase in $[Ca^{2+}]_i$, to activation of PLA_2, to arachidonic acid release, to generation of toxigenic eicosanoids. With the discovery that TXB_2 can cause bleb formation and cell death [26, 27], it was postulated [27] that if TXB_2 is acting by stimulation of proteolysis of cytoskeletal proteins, then TXB_2-dependent bleb formation should be inhibited by leupeptin. Experiments confirming the suppositon showed that TXB_2-dependent hepatocyte bleb formation was markedly inhibited by leupeptin.

Disturbance of a PGI_2-TXA_2,TXB_2 balance is probably involved in other instances of toxigenic cell injury. For example, in rat peritoneal leucocytes incubated with CCl_4, PLA_2 is activated and TXB_2, with lesser amounts of PGE_2, 6-keto-$PGF_{1\alpha}$, and LTB_4, are released [35]. In other work [39], elevation of TXB_2 in the liver was observed 6 h after injection of CCl_4 into mice. Administration of either an inhibitor of TXA_2 synthesis or a TXA_2 receptor antagonist attenuated the CCl_4-dependent liver injury. Also, a stable TXA_2, mimetic elevated serum GOT and GPT levels, and produced histologically detectable liver cell damage.

In conclusion, it is clear from this review that dense mysteries surround the general problem of the mechanisms at work in toxigenic liver cell injury, especially with respect to the secondary mechanisms that ultimately cause death of the cell. However, the recent work on eicosanoid involvement has opened up new perspectives, not only for study of the molecular pathology of liver cell injury and necrosis, but also for study of the molecular mechanisms involved in prostaglandin cytoprotection and thromboxane cytotoxicity.

Acknowledgement. The original work reported in this review was supported by a grant (No. ES 01821) from the National Institute of Environmental Health Sciences, National Institutes of Health.

References

1. Albano E, Carini R, Parola M, Bellomo G, Goria-Gatti L, Poli G, Dianzani MU (1989) Effects of carbon tetrachloride on calcium homeostasis. A critical reconsideration. Biochem Pharmacol 38:2719–2725
2. Araki H, Lefer AM (1980) Cytoprotective actions of prostacyclin during hypoxia in the isolated perfused rat liver. Am J Physiol 238 (Heart Circ Physiol 7):H176–H181
3. Bellomo G, Orrenius S (1985) Altered thiol and calcium homeostasis in oxidative hepatocellular injury. Hepatology 5:876–882
4. Berger ML, Bhatt H, Combes B, Estabrook RW (1986) CCl$_4$-induced toxicity in isolated hepatocytes: the importance of direct solvent injury. Hepatology 6:36–45
5. Britton RS, Bacon BR, and Recknagel RO (1987) Lipid peroxidation and associated organelle dysfunction in iron overload. Chem Phys Lipids 45:207–239
6. Chiarpotto E, Biasi F, Comoglio A, Leonarduzzi G, Poli G, and Dianzani MU (1990) CCl$_4$-induced increase of hepatocyte free arachidonate level: pathogenesis and contribution to cell death. Chem Biol Interact 74:195–206
7. Chien KR, Abrams J, Serroni A, Martin JR, Farber JL (1978) Accelerated phospholipid degradation and associated membrane dysfunction in irreversible, ischemic liver cell injury. J Biol Chem 253:4809–4817
8. Comporti M (1985) Biology of disease: lipid peroxidation and cellular damage in toxic liver injury. Lab Invest 53:599–623
9. Comporti M (1987) Glutathione depleting agents and lipid peroxidation. Chem Phys Lipids 45:143–169
10. Decker K (1985) Eicosanoids, signal molecules of liver cells. Semin Liver Dis 5:175–190
11. Diniz CR, Corrado AP (1971) Venoms of insects and arachnids. International Encyclopedia of Pharmacology and Therapeutics, section 71, vol 2. Pergamon, New York, pp 117–140
12. Divald A, Ujhelyi A, Jeney A, Lapis J, and Institoris L (1985) Hepatoprotective effects of prostacylins on CCl$_4$-induced liver injury in rats. Exp Mol Pathol 42:163–166
13. Dolak JA, Waller RL, Glende EA Jr, Recknagel RO (1988) Liver cell calcium homeostasis in carbon tetrachloride liver cell injury: new data with fura-2. J Biochem Toxicol 3:329–342
14. Dutera SM, Byrne WL, and Ganoza MC (1968) Studies on the phospholipid requirement of glucose-6-phosphatase. J Biol Chem 243:2216–2228
15. Farber JL (1982) Biology of disease: membrane injury and calcium homeostasis in the pathogenesis of coagulative necrosis. Lab Invest 47:114–123

16. Feher J, Csomos G, Vareckei A (1987) Free radical reactions in medicine. Springer, Berlin Heidelberg New York
17. Funck-Brentano C, Tinel M, DeGoot C, Letteron P, Babany G, Pessayre D (1984) Protective effect of 16,16-dimethyl prostaglandin E_2 on the hepatotoxicity of bromobenzene in mice. Biochem Pharmacol 33:89–96
18. Glende EA Jr, Pushpendran CK (1986) Activation of phospholipase A_2 by carbon tetrachloride in isolated rat hepatocytes. Biochem Pharmacol 35:3301–3307
19. Gores G, Herman B, Lemasters JJ (1990) Plasma membrane bleb formation and rupture: a common feature of hepatocellular injury. Hepatology 11:690–698
20. Guarner F, Fremont-Smith M, Corzo J, Quiroga J, Rodriquez JL, Prieto J (1983) In vivo and in vitro effects of arachidonic acid products on liver cell integrity following carbon tetrachloride poisoning. Adv Prostaglandin Thromboxane Leukotriene Res 12:75–82
21. Guarner F, Fremont-Smith M, Prieto J (1985) Cytoprotective effects of prostaglandins on isolated liver cells. Liver 5:35–90
22. Guarner F, Boughton-Smith NK, Blackwell GJ, Moncada S (1988) Reduction by prostacylcin of acetaminophen-induced liver toxicity in the mouse. Hepatology 8:248–253
23. Henriques SB, Henriques OB (1971) Pharmacology and toxicology of snake venoms. International Encyclopedia of Pharmacological Therapeutics, section 71, vol 1. Pergamon, New York, pp 215–368
24. Horton AA, Fairhurst S (1987) Lipid peroxidation and mechanisms of toxicity. CRC Crit Rev Toxicol 18:27–79
25. Horton AA, Wood JM (1989a) Effect of inhibitors of phospholipase A_2, cyclooxygenase and thromboxane synthetase on paracetamol hepatotoxicity in the rat. Eicosanoids 2:123–129
26. Horton AA, Wood JM (1989b) Prevention of Ca^{2+}-induced hepatocyte plasma membrane bleb formation by inhibitors of eicosanoid synthesis. J Lipid Mediators 1:231–242
27. Horton AA, Wood JM (1990) Prevention of thromboxane B_2-induced hepatocyte plasma membrane bleb formation by certain prostaglandins and a proteinase inhibitor. Biochim Biophys Acta 1022:319–324
28. Keppler D, Hagmann W, Rapp S, Denzlinger C, Koch HK (1985) The relation of leukotrienes to liver injury. Hepatology 5:883–891
29. Koch RR, Glende EA Jr, Recknagel RO (1974) Hepatotoxicity of bromotrichloromethane: bond dissociation energy and lipoperoxidation. Biochem Pharmacol 23:2907–2915
30. Lemasters JJ, diGuiseppi J, Nieminen AL, Herman B (1987) Blebbing, free Ca^{2+} and mitochondrial membrane potential preceding cell death in hepatocytes. Nature 325:78–81
31. Long RM, Moore L (1986) Elevated cytosolic calcium in rat hepatocytes exposed to carbon tetrachloride. J Pharmacol Exp Ther 238:186–191
32. Long RM, Moore L (1987) Cytosolic calcium after carbon tetrachloride, 1,1-dichloroethylene, and phenylephrine exposure. Studies in rat hepatocytes with phosphorylase a and quin-2. Biochem Pharmacol 36:1215–1221
33. Lowrey K, Glende EA Jr, Recknagel RO (1981a) Destruction of liver microsomal calcium pump activity by carbon tetrachloride and bromotrichloromethane. Biochem Pharmacol 30:135–140
34. Lowrey K, Glende EA Jr, Recknagel RO (1981b) Rapid depression of rat liver microsomal calcium pump activity after administration of carbon tetrachloride or bromotrichloromethane and lack of effect after ethanol. Toxicol Appl Pharmacol 59:389–394
35. Lynch TJ, Blackwell GJ, Moncada S (1985) Carbon tetrachloride-induced eicosanoid synthesis and release from rat peritoneal leucocytes. Biochem Pharmacol 9:1515–1521
36. Marinovich M, Flaminio LM, Papagni M, Galli CL (1987) Evaluation of the cytoprotective effect of natural and synthetic prostaglandins in CCl_4-induced liver cell damage. Adv Prostaglandin Thromboxane Leucotriene Res 17:1094–1097

37. Moore L, Davenport GR, Landon EJ (1976) Calcium uptake by a rat liver microsomal subcellular fraction in response to in vivo administration of carbon tetrachloride. J Biol Chem 251:1197–1201

38. Nagai H, Shimazawa T, Yakuo I, Aoki M, Koda A, Kasahara M (1989a) Role of peptide-leukotrienes in liver injury in mice. Inflammation 13:673–680

39. Nagai H, Shimazawa T, Yakuo I, Aoki M, Koda A, Kasahara M (1989b) The role of thromboxane A$_2$ [TXA$_2$] in liver injury in mice. Prostglandins 38:439–446

40. Nicotera P, Hartzell P, Baldi C, Svenson, S-A, Bellomo G, Orrenius S (1986a) Cystamine induces toxicity in hepatocytes through the elevation of cytosolic Ca^{2+} and the stimulation of a nonlysosomal proteolytic system. J Biol Chem 261:14628–14635

41. Nicotera P, Hartzell P, Davis G, Orrenius S (1986b) The formation of plasma membrane blebs in hepatocytes exposed to agents that increase cytosolic Ca^{2+} is mediated by activation of a non-lysosomal proteolytic system. FEBS Lett 209:139–144

42. Parthasarathy S, Steinbrecher UP, Barnett J, Witztum JL, Steinberg D (1985) Essential role of phospholipase A$_2$ activity in endothelial cell-induced modification of low density lipoproteins. Proc Natl Acad Sci USA 82:3000–3004

43. Recknagel RO (1983) Minireview: a new direction in the study of carbon tetrachloride hepatotoxicity. Life Sci 33:401–408

44. Recknagel RO, Glende EA Jr (1989) The carbon tetrachloride hepatotoxicity model: free radicals and calcium homeostasis. In: Quintanila AT, Weber H (eds) Free radicals and antioxidants in biomedicine, vol 3. CRC Boca Raton, pp 3–16

45. Recknagel RO, Glende EA Jr, Hruszkewycz AM (1977) Chemical mechanisms in carbon tetrachloride toxicity. In: Pryor WA (ed) Free radicals in biology, vol 3. Academic, New York, pp 97–132

46. Recknagel RO, Glende EA Jr, Dolak JA, Waller RL (1989) Mechanisms of carbon tetrachloride toxicity. Pharmacol Ther 43:139–154

47. Recknagel RO, Glende EA Jr, Britton RS (1991) Free radical damage and lipid peroxidation. In: Meeks RG, Harrison SD, Bull RJ (eds) Hepatotoxicology. CRC, Boca Raton, FLA, pp 395–430

48. Reed DJ (1990) Glutathione: toxicological implications. Ann Rev Pharmcol Toxicol 30:603–631

49. Robert A (1976) Antisecretory, antiulcer, cytoprotective and diarrheogenic properties of prostaglandins. In: Samuelsson B, Paoletti R (eds) Advances in Prostaglandin and Thromboxane Research, vol 2. Raven, New York, pp 507–520

50. Robert A (1981) Current history of cytoprotection. Prostaglandins 21 (Suppl):89–96

51. Rush B, Merritt MV, Kaluzny M, vanSchoick T, Brunden MN, Ruwart MJ (1986) Studies on the mechanism of the protective action of 16,16-dimethyl PGE$_2$ in carbon tetrachloride induced acute hepatic injury in the rat. Prostaglandins 32:439–455

52. Rush BD, Wilkinson KF, Nichols NM, Ochoa R, Brunden MN, Ruwart MJ (1989) Hepatic protection by 16,16-dimethyl PGE$_2$ (DMPG) against acute aflatoxin B$_1$-induced injury in the rat. Prostaglandins 37:683–692

53. Ruwart MJ, Kolaja GJ, Friedle NM, Rush BD, Tarnawski A, Stachura J, Mach T, Ivey KJ (1981a) Protection against CCl$_4$-induced liver cell damage by 16,16-dimethyl PGE$_2$. Gastroenterology 80:1266

54. Ruwart MJ, Rush BD, Friedle NM, Piper RC, Kolaja GJ (1981b) Protective effects of 16,16-dimethyl PGE$_2$ on the liver and kidney. Prostaglandins 21:97–102

55. Ruwart MJ, Friedle NM, Rush BD (1982a) 16,16-Dimethyl PGE$_2$ protects in vitro from carbon tetrachloride-induced damage. Gastroenterology 82:1166

56. Ruwart MJ, Rush BD, Friedle NM (1982b) 16,16-Dimethyl PGE$_2$ partially prevents necrosis due to aflatoxin in rats. Gastroenterology 82:1167

57. Ruwart MJ, Rush BD, Friedle NM, Stachura J, Tarnawski A (1984) 16,16-Dimethyl PGE$_2$ protection against α-naphthylisothiocyanate-induced experimental cholangitis in the rat. Hepatology 4:658–660

58. Ruwart MJ, Nichols NM, Hedeen J, Rush BD, Stachura J (1985) 16,16-Dimethyl PGE$_2$ and fatty acids protect hepatocytes against CCl$_4$-induced damage. In vitro Cell Dev Biol 21:450–452

59. Sakagami Y, Mizoguchi Y, Saki S, Kobayashi K, Morisawa S, Yamamoto S (1988) Release of peptide leukotrienes from rat Kupffer cells. Biochem Biophys Res Commun 156:217–221

60. Sevanian A, Kim E (1985) Phospholipase A_2 dependent release of fatty acids from peroxidized membranes. J Free Radic Biol Med 1:263–271

61. Shier WT, duBourdieu JJ (1982) Role of phospholipid hydrolysis in the mechanism of toxic cell death by calcium and ionophore A23187. Biochem Biophys Res Commun 109:106–112

62. Shimazawa T, Nagai H, Koda A, Kasahara M (1990) The effects of thromboxane A_2 inhibitors (OKY-046 and ONO-3708) and leukotriene inhibitors (AA-861 and LY-171883) on CCl_4-induced chronic liver injury in mice. Prostaglandins Leukot Essent Fatty Acids 40:67–71

63. Slater TE, Sawyer BC (1971) The stimulatory effects of carbon tetrachloride on peroxidative reactions in rat liver fractions in vitro. Interaction sites in the endoplasmic reticulum. Biochem J 123:815–821

64. Stacey NH, Klaassen CD (1982) Effects of phospholipase A_2 inhibitors on diethylmaleate-induced lipid peroxidation and cellular injury in isolated hepatocytes. Toxicol Environ Health 9:439–450

65. Stachura J, Tarnawski A, Ivey KJ, Mach T, Bogdal J, Szczudrawa J, Klimczyk B (1981a) Prostaglandin protection of carbon tetrachloride-induced liver cell necrosis in the rat. Gastroenterology 81:211–217

66. Stachura J, Tarnawski A, Ivey JJ, Ruwart MJ, Ruch BD, Friedle NM, Szczudrawa J, Mach T (1981b) 16,16-Dimethyl PGE_2 protection of rat liver against acute injury by galactosamine, acetaminophen, ethanol and ANIT. Gastroenterology 80:1349

67. Tomasi A, Albano E, Banni S, Botti B, Corongiu F, Dessi MA, Iannone A, Vannini V, Dianzani MU (1987) Free-radical metabolism of carbon tetrachloride in rat liver mitochondria: a study of the mechanism of action. Biochem J 246:313–317

68. Trudell JR, Bendix M, Bosterling B (1984) Hypoxia potentiates killing of hepatocyte monolayers by leukotrienes, hydroperoxyeicosatetraenoic acids, or calcium ionophore A23187. Biochim Biophys Acta 803:338–341

69. Tsokos-Kuhn JO (1988) Lethal injury by diquat redox cycling in an isolated hepatocyte model. Arch Biochem Biophys 265:415–424

70. Ungemach FR (1985) Plasma membrane damage of hepatocytes following lipid peroxidation: involvement of phospholipase A_2. In: Poli G, Cheeseman KH, Dianzani MU, Slater TF (eds) Free radicals in liver injury. IRL, Oxford, pp 127–134

71. Ungemach FR (1987) Pathobiological mechanisms of hepatocellular damage following lipid peroxidation. Chem Phys Lipids 45:171–205

72. Ungemach FR (1989) Prevention of liver cell damage following lipid peroxidation by depression of lysophosphatide formation. Arch Toxicol (Suppl) 13:275–281

73. vanKuijk FJGM, Sevanian A, Handelman GJ, Dratz EA (1987) A new role for phospholipase A_2: protection of membranes from lipid peroxidation. Trends Biochem Sci 12:31–34

74. Weglicki WB, Dickens BF, Tong Mak I (1984) Enhanced lysosomal phospholipid degradation and lysophospholipid production due to free radicals. Biochem Biophys Res Commun 124:229–235

75. Yasuda M, Fujita T (1977) Effect of lipid peroxidation on phospholipase A_2 activity of rat liver mitochondria. Jpn J Pharmacol 27:429–435

76. Zimmerman H (1978) Hepatotoxicity. Appleton-Century-Crofts, New York

Lipid Peroxidation and Associated Hepatic Organelle Dysfunction in Iron Overload

B.R. Bacon, R.S. Britton, and R. O'Neill

Introduction

It has recently been recognized that genetic hemochromatosis (GH) is one of the most common inherited disorders. Approximately 0.1%–0.8% of whites of northern European ancestry are affected and the frequency of heterozygotes is thought to be about 5%–16% of the population [7, 12, 17, 20, 24, 34, 39, 43, 44, 53, 54]. In GH, there is an inappropriate increase in intestinal absorption of iron leading to progressive deposition of excess iron in parenchymal cells of the liver and several other organs [26, 28, 41, 47, 56]. The liver is the major recipient of the excess absorbed iron and after several years of high tissue iron concentrations, fibrosis and eventually cirrhosis develop [8, 38, 42, 56]. In addition, excess iron deposition in the liver is found in a variety of other disorders leading to secondary iron overload (e.g., thalassemia, chronic liver disease, porphyria cutanea tarda, sideroblastic anemia, African iron overload) [8, 26, 38, 41, 56].

Clinical evidence for hepatotoxicity caused by excess iron is now well established and has been provided by studies of patients with GH, African iron overload, and secondary hemochromatosis caused by β-thalassemia. In these conditions a correlation has been demonstrated between the hepatic iron concentration and the occurrence of liver injury, and therapeutic reduction of hepatic iron by either phlebotomy or chelation therapy has resulted in clinical improvement [6, 9, 10, 19, 32, 42, 48]. In addition to the known association of chronic iron overload and the development of hepatic fibrosis and cirrhosis, there is now sufficient clinical evidence to indicate an increased risk of hepatocellular carcinoma in hemochromatotic patients with cirrhosis [42]. Despite this convincing clinical evidence for liver injury as a result of excess iron, the specific pathophysiological mechanisms for hepatocellular injury, fibrosis, cirrhosis, and carcinoma in iron overload are poorly understood. Figure 1 outlines several proposed pathways by which excess iron causes hepatic damage. Iron-induced oxidative injury to membrane phospholipids and to DNA may be a unifying mechanism underlying the several major pathways of cellular injury in iron overload. It is well known from in vitro experiments that ionic iron (usually with a reductant present) stimulates lipid peroxidation and causes oxidative damage to isolated cells, organelles, and DNA [13]. This paper will review the evidence

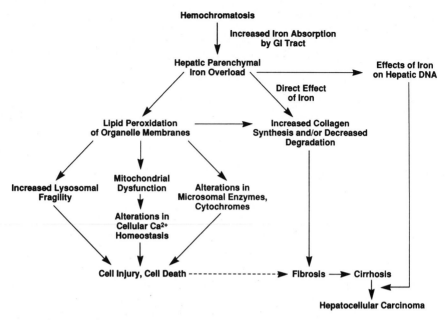

Fig. 1. Proposed pathophysiological mechanisms of cellular injury and hepatic fibrosis in iron overload

that iron overload in vivo results in lipid peroxidation and damage to hepatocytes, hepatocellular organelles, and hepatic DNA.

Iron-Induced Lipid Peroxidation

Very little information exists regarding the occurrence of lipid peroxidation in humans with any of the clinical syndromes of hemochromatosis. Accordingly, most of our understanding of mechanisms of cellular injury related to lipid peroxidation in iron overload has come from studies in experimental animals. Dillard and Tappel [22] found an increase in ethane and pentane exhalation in rats that received multiple injections of iron-dextran and this effect was exacerbated by α-tocopherol deficiency [23]. While these experiments demonstrated that large parenteral doses of an iron chelate can result in increased lipid peroxidation, the site of peroxidation (organ or organelle) was unknown.

Lipid peroxidation can also be measured using the thiobarbituric acid (TBA) test. In 1962, Golberg et al. [25] demonstrated increased levels of TBA reactants in liver, kidney, muscle, and skin taken from mice and rats that had received several intramuscular injections of iron-dextran. Hultcrantz et al. [31] found increased amounts of TBA reactants in the kidneys and liver of rats receiving multiple intramuscular injections of iron-sorbitol-citric acid and we recently demonstrated an increased level of hepatic TBA reactants in rats fed a diet supplemented with carbonyl (finely divided elemental) iron [15].

Measurement of TBA reactants from intact tissues does not allow determination of the specific subcellular sites of iron-induced lipid peroxidation and may be complicated by in vitro peroxidation during the assay procedure [13]. One method which abrogates these difficulties is measurement of conjugated dienes in phospholipids extracted from subcellular organelles prepared by routine differential centrifugation. We have used the conjugated diene assay to determine whether lipid peroxidation occurred in vivo in hepatic mitochondrial and microsomal fractions prepared from rats with experimental iron overload [1–5]. Iron overload was produced either by daily intraperitoneal injections of ferric nitrilotriacetate or by dietary supplementation with carbonyl iron. Evidence of mitochondrial lipid peroxidation in vivo was demonstrated in both models of experimental chronic iron overload. Increased conjugated diene formation was detected in microsomal lipids only at the higher liver iron concentrations achieved by dietary carbonyl iron supplementation [2]. In subsequent experiments, we concluded that iron-induced peroxidative damage to hepatic membrane lipids in chronic iron overload depends on the hepatic iron concentration, and the critical "threshold" level of iron required is higher for microsomes than for mitochondria [1].

In summary, a number of investigators have found evidence of increased lipid peroxidation in experimental iron overload using various methods for administering iron and for detecting the presence of lipid peroxidation. Subsequent studies have been performed to determine the impact of this peroxidation on membrane-dependent functions in hepatic organelles.

Hepatic Organelle Dysfunction in Experimental Iron Overload

Lysosomes. Experimental iron overload reduces latent lysosomal activity in liver samples, suggesting that lysosomal fragility is increased [30, 33, 46, 49]. Using rats fed a diet supplemented with carbonyl iron, LeSage et al. [33] observed that latent lysosomal enzyme activity was decreased only after a threshold hepatic iron concentration of approximately 2400 µg/g wet wt. was exceeded. In this model of experimental iron overload, there was an increase in the total activity of lysosomal enzymes in liver homogenates, similar to the changes in liver biopsy specimens from iron-loaded patients reported by Seymour and Peters [50]. In a recent report using the carbonyl iron model, Myers et al. [40] have shown evidence of increased lipid peroxidation in lysosomal membranes (TBA reactants) with an associated decrease in membrane fluidity and a defect in lysosomal acidification (increased lysosomal pH).

Microsomes. Several investigators have reported decreases in hepatic cytochrome P-450 concentration and in the activity of some of the microsomal drug-metabolizing enzyme systems using models of either acute or chronic parenteral iron loading [11, 21, 35]. When dietary carbonyl iron supplementation was used as a method for producing experimental iron overload, evidence of hepatic microsomal lipid peroxidation was found at high hepatic iron concentrations in association with decreases in both the concentration of cytochromes P-450 and b_5 and in the activity of aminopyrine demethylase [4]. From these experiments it was concluded that iron-induced lipid peroxidation to microsomal membranes was probably responsible for the decrease in cytochrome levels and drug-metabolizing capacity. Recent experiments from our laboratory have shown that calcium sequestration by hepatic microsomes is depressed in experimental iron overload [16] (see Fig. 2). This could result in impaired calcium homeostasis in the iron-loaded liver.

Mitochondria. As one measure of mitochondrial integrity, the activity of the respiratory chain has been evaluated in hepatic mitochondria isolated from iron-loaded rats. In normal mitochondria, electron transport is tightly coupled to the phosphorylation of ADP; thus, in the absence of ADP (state 4), the rate of oxygen consumption is low, while with ADP present (state 3) the rate of oxygen consumption is increased. Therefore, normal mitochondria have a high respiratory control ratio, which is calculated by dividing the state 3 respiratory rate by the state 4 respiratory rate. We have observed significant decreases in the state 3 respiratory rates and in the respiratory control ratio for three substrates (glutamate, β-hydroxybutyrate, and succinate) at moderate degrees of chronic hepatic iron overload in vivo, produced by dietary supplementation with carbonyl iron [3]. This functional impairment became evident at hepatic non-heme iron concentrations at which mitochondrial lipid peroxidation (conjugated dienes) occurred. From these studies, we concluded that moderate degrees of chronic hepatic iron overload in vivo result in an inhibitory defect in the mitochondrial electron transport chain as evidenced by a decrease in state 3 respiration. Since the electron transport chain is embedded in the inner mitochondrial membrane, it seems possible that a change in membrane properties induced by lipid peroxidation might be responsible for the observed defects. Masini and coworkers [36, 37] have reported that mitochondria from iron-loaded rats have a significant reduction in the transmembrane potential and lower levels of potassium.

Because α-tocopherol plays an important role in protecting membrane phospholipids from peroxidative decomposition [55], experiments were designed to determine the effect of α-tocopherol supplementation on lipid peroxidation and on the altered mitochondrial function seen in experimental dietary iron overload [5]. In these experiments, parenteral α-tocopherol was administered throughout the feeding period to produce iron-loaded rats

A.

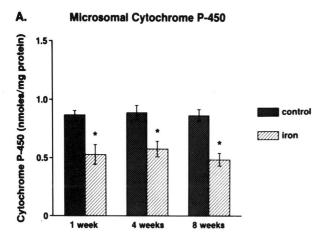

B.

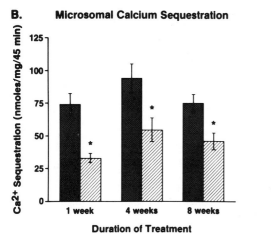

Fig. 2A,B. Effects of chronic dietary iron overload on **A** hepatic microsomal cytochrome P-450 levels and **B** microsomal calcium sequestration. Experimental iron overload was produced by feeding three groups of rats a chow diet supplemented with 3% (w/w) carbonyl iron for 1, 4, or 8 weeks to achieve different degrees of hepatic iron overload. Control rats received the chow diet alone. Hepatic microsomes were isolated, and cytochrome P-450 levels were determined. Microsomal calcium sequestration was measured using a filtration assay after a 45-min incubation in a buffer containing $^{45}Ca^{2+}$. Values are means $\pm$SEM ($n = 7$–10). Statistically significant differences are indicated by *an asterisk* ($p < 0.01$) [16]

with elevated (threefold) hepatic α-tocopherol levels: comparisons were made with iron-loaded rats with normal or deficient α-tocopherol levels. α-Tocopherol deficiency alone (in the absence of iron overload) did not result in mitochondrial lipid peroxidation or alterations in mitochondrial function. Significant reductions in the mitochondrial respiratory control ratio and oxidative phosphorylation (ADP/O ratios) were seen in association with increased conjugated dienes in all iron-loaded rats regardless

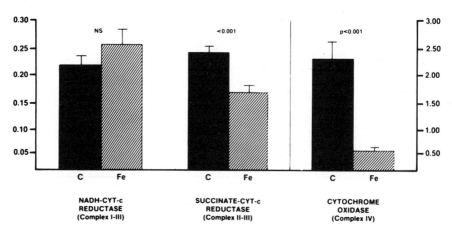

Fig. 3. Effect of chronic dietary iron overload on hepatic mitochondrial oxidases and reductases. Experimental iron overload was produced by feeding rats ($n = 6$) a chow diet supplemented with 3.0% carbonyl iron for 7–8 weeks achieving a mean ($\pm$SEM) hepatic iron concentration of $2510 \pm 470\,\mu g/g$. Control rats ($n = 6$) received chow diet alone and had a mean ($\pm$SEM) hepatic iron concentration of $150 \pm 49\,\mu g/g$. There was a significant reduction in succinate-cytochrome-c-reductase activity and a marked decrease in cytochrome oxidase activity in hepatic mitochondria isolated from rats with iron overload

of the α-tocopherol status (deficient, normal, or increased). Thus, the role of α-tocopherol supplementation in potentially decreasing the toxic effects of iron, both in experimental studies and in clinical practice, is still unclear.

Experiments have also been performed in which the activity of various mitochondrial oxidases and reductases has been examined in dietary carbonyl iron overload. At moderate degrees of hepatic iron overload, no changes in NADH-cytochrome-c-reductase activity (complex I–III) existed; however, significant decreases in succinate-cytochrome-c-reductase (complex II–III) and marked (70%) decreases in cytochrome oxidase activity were found (complex IV) (Fig. 3). Cytochrome oxidase is functionally dependent on intact cardiolipin, a phospholipid unique to mitochondria. Since cardiolipin contains a high percentage of polyunsaturated fatty acids, it may be susceptible to iron-induced peroxidative damage and this, in turn, could decrease cytochrome oxidase activity.

In order to examine the consequences of these alterations in mitochondrial function on the hepatic energy state, the concentrations of adenine nucleotides (ATP, ADP, AMP) were measured in the livers of rats with dietary iron overload. At high hepatic iron concentrations at which mitochondrial substrate oxidation was significantly reduced, there

NUCLEOTIDES
μmoles/gm liver

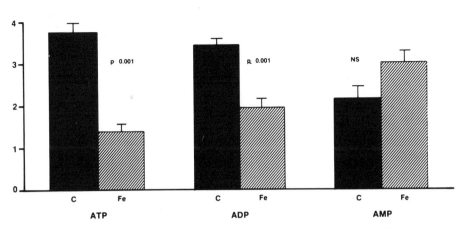

Fig. 4. Effect of chronic dietary iron overload on hepatic nucleotide levels. Experimental iron overload was produced by feeding rats ($n = 12$) a chow diet supplemented with 3.0%–3.3% carbonyl iron for 8 weeks achieving a mean (±SEM) hepatic iron concentration of 4635 ± 398 μg/g. Control rats ($n = 12$) received the chow diet alone and had a mean (±SEM) hepatic iron concentration of 210 ± 80 μg/g. There were significant reductions in levels of hepatic ATP and ADP in iron overload

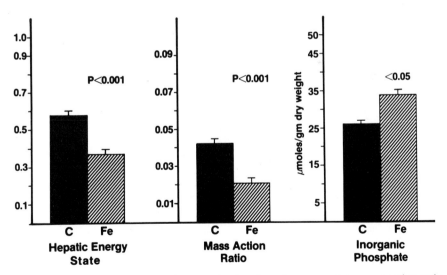

Fig. 5. Effect of chronic dietary iron overload on hepatic energy state, mass action ratio, and inorganic phosphate levels. Rats were treated as described in Fig. 4. The energy state is defined as $[(ATP) + 0.5(ADP)]/[(ATP) + (ADP) + (AMP)]$, and the mass action ratio as $[ATP]/[ADP][P_i]$. These was a significant decrease in the hepatic energy state and mass action ratio in iron overload. The inorganic phosphate level was increased in iron overload

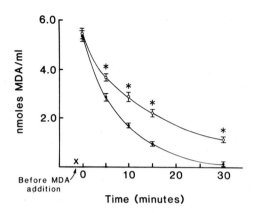

Fig. 6. Mitochondrial malondialdehyde (*MDA*) metabolism in iron overload. Mitochondria (2 mg/ml) were prepared from iron-loaded, ○–○, and control, ●–●, rats and incubated in a buffer containing 6 μM MDA and 12 mM ATP for 30 min at 37°C. Aliquots were taken for assay of MDA at −1, 0, 5, 10, 15, and 30 min. At 5, 10, 15, and 30 min there were significant reductions (*, $p < 0.05$ at 5 min; $p < 0.01$ at 10, 15, and 30 min) in the amount of MDA metabolized by mitochondria from iron-loaded rats compared to controls (mean ±SEM, $n = 6$) [15]

was a 60% reduction in hepatic ATP concentration with a corresponding decrease in the hepatic energy state and mass action ratio (Figs. 4, 5).

Another consequence of mitochondrial dysfunction in iron overload is impaired metabolism of aldehydic by-products of lipid peroxidation. We have recently shown a significant decrease in the rate of malondialdehyde (MDA) metabolism by mitochondria from iron-loaded rats [15] (Fig. 6). This decrease may be due to reductions in both the oxidative capacity of the mitochondria and the activity of mitochondrial aldehyde dehydrogenase. Thus, iron-induced mitochondrial lipid peroxidation not only causes an increase in MDA production but also results in impaired metabolism of MDA.

Nucleic Acids. One of the most common causes of death in patients with long-standing GH is the development of hepatocellular carcinoma [42]. Accordingly, it has been postulated that iron-induced oxidative damage to nucleic acids may play an important role in this particular form of liver injury. Shires [52] has demonstrated DNA strand breaks in rat hepatic nuclei subjected to iron-induced oxidation in vitro. Similarly, when rat liver mitochondria are peroxidized in vitro, there is damage to mitochondrial DNA [29]. In preliminary experiments from our laboratory we have found that chronic iron overload damages hepatic DNA in vivo. In these studies, DNA strand breaks were measured in liver samples using an alkali-unwinding assay. Much additional work will be necessary to relate these findings of DNA damage to ultimate carcinogenesis.

Experiments in Isolated Iron-Loaded Hepatocytes

A new system has been developed to investigate the mechanisms of hepatocellular injury in chronic iron overload in which isolated iron-loaded hepatocytes are incubated in suspension [51]. In this system, cell viability was significantly reduced at 3 and 4 h in iron-loaded hepatocytes compared to controls, and this decrease in viability was preceded by an increase in iron-dependent lipid peroxidation. Extensive degenerative ultrastructural changes were observed in iron-loaded hepatocytes compared to controls after 4 h of incubation. In vitro iron chelation with either deferoxamine or apotransferrin protected against lipid peroxidation, loss of viability, and ultrastructural damage in iron-loaded hepatocytes. In addition, exogenous α-tocopherol had a protective effect in these cells. The protective action of iron chelators and antioxidants supports a strong association between iron-dependent lipid peroxidation and hepatocellular injury in isolated iron-loaded hepatocytes.

Having demonstrated the usefulness of the isolated iron-loaded hepatocyte system for studying the effects of chronic iron overload on cellular injury, we next sought to perform studies in which an association between altered mitochondrial calcium cycling and decreases in cell viability could be established. Ruthenium red is known to be a specific inhibitor of the mitochondrial calcium uniport, can enter hepatocytes, and is not an antioxidant [57]. In experiments by Thomas and Reed [57], addition of ruthenium red to isolated hepatocytes, which were oxidatively stressed by incubation in a calcium-deficient buffer, preserved cell viability and protected against lipid peroxidation and loss of glutathione. Similarly, Gores and colleagues have shown that ruthenium red decreased the effects of t-butyl hydroperoxide on the mitochondrial transmembrane potential and cell viability in isolated hepatocytes [27, personal communication]. Therefore, we examined if ruthenium red added to isolated iron-loaded hepatocytes would protect against loss of cell viability by blocking mitochondrial calcium cycling. Isolated iron-loaded hepatocytes were prepared from rats with chronic iron overload and their respective controls. Ruthenium red $(50 \mu M)$ was added at time 0. Aliquots of hepatocyte suspensions were taken at 0, 1, 2, 3, and 4 h and tested for cell injury using trypan blue exclusion and the release of lactate dehydrogenase (LDH). Ruthenium red had no effect on the viability of control hepatocytes. However, addition of ruthenium red maintained the viability of iron-loaded hepatocytes at control levels. These results are summarized in Figs. 7 and 8.

Summary and Conclusions

In experimental animals, iron overload clearly can cause lipid peroxidation in vivo, as evidenced by increased hepatic conjugated dienes, TBA re-

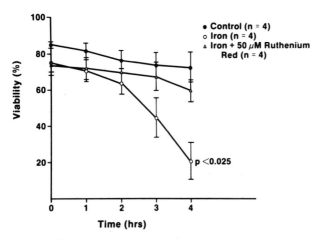

Fig. 7. Effect of ruthenium red on the viability of isolated iron-loaded hepatocytes. Exprimental iron overload was achieved by feeding rats ($n = 4$) a diet supplemented with 3% carbonyl iron achieving a mean (±SEM) hepatic iron concentration of $4215 \pm 880\,\mu g/g$. Control rats ($n = 4$) received a chow diet alone and had a mean (±SEM) hepatic iron concentration of $265 \pm 11\,\mu g/g$. Isolated hepatocytes were prepared and incubated as described in [51]. Viability was assessed by trypan blue exclusion. Addition of ruthenium red maintained viability in the iron-loaded hepatocytes at control levels

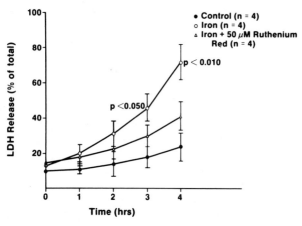

Fig. 8. Effect of ruthenium red on lactate dehydrogenase (*LDH*) release in isolated iron-loaded hepatocytes. Rats were treated as described in Fig. 7. Addition of ruthenium red maintained LDH release in iron-loaded hepatocytes at control levels

actants, and alkane exhalation. For peroxidation to occur, the iron concentration must exceed a certain threshold level. Presumably, once this threshold is reached, the capacity of the hepatocyte to maintain iron in nontoxic storage forms is exceeded, with a subsequent increase in a catalytically active pool of low-molecular-weight iron that is responsible for the production of free radicals and subsequent lipid peroxidation. Recent experiments from our laboratory have confirmed an increase in such a pool of iron in chronic dietary iron overload [14]. In experimental iron overload, a number of studies have described abnormalities in hepatic organelles (lysosomes, mitochondria, microsomes) in association with lipid peroxidation. Although the association between these functional abnormalities and lipid peroxidation does not prove direct causality, it seems likely that lipid peroxidation is at least partially responsible because similar functional defects are produced in these organelles in vitro by iron-induced peroxidation. Furthermore, experiments using isolated iron-loaded hepatocytes have confirmed the causal relationship between iron-induced lipid peroxidation and cell death. Reduced cellular ATP levels, damage to DNA, and impaired cellular calcium homeostasis may all contribute to hepatocellular injury in iron overload.

The question as to whether organelle lipid peroxidation and associated dysfunction are involved in the development of hepatic fibrosis and cirrhosis is as yet unanswered. However, recent experiments by Chojkier, Brenner, and colleagues have shown that iron-induced lipid peroxidation causes an increase in collagen gene transcription and collagen production in cultured human fibroblasts, suggesting that lipid peroxidation may in fact play an important role in fibrogenesis [18]. The demonstration that hepatic fibrosis is produced in rats with long-term dietary iron overload [45] will allow this model to be utilized to study the relationships between lipid peroxidation, hepatocellular injury, and fibrosis.

References

1. Bacon BR, Brittenham GM, Tavill AS, McLaren CE, Park CH, Recknagel RO (1983a) Hepatic lipid peroxidation in vivo in rats with chronic dietary iron overload is dependent on hepatic iron concentration. Trans Assoc Am Physicians 96:146–154
2. Bacon BR, Tavill AS, Brittenham GM, Park CH, Recknagel RO (1983b) Hepatic lipid peroxidation in vivo in rats with chronic iron overload. J Clin Invest 71:429–439
3. Bacon BR, Park CH, Brittenham GM, O'Neill R, Tavill AS (1985) Hepatic mitochondrial oxidative metabolism in rats with chronic dietary iron overload. Hepatology 5:789–797
4. Bacon BR, Healey JF, Brittenham GM, Park CH, Nunnari J, Tavill AS, Bonkovsky HL (1986) Hepatic microsomal function in rats with chronic dietary iron overload. Gastroenterology 90:1844–1853
5. Bacon BR, Britton RS, O'Neill R (1989) Effects of vitamin E deficiency on hepatic mitochondrial lipid peroxidation and oxidative metabolism in rats with chronic dietary iron overload. Hepatology 9:398–404
6. Barry M, Flynn DM, Letsky EA, Risdon RA (1974) Long-term chelation therapy in thalassemia major: effect on liver iron concentration, liver histology and clinical progress. Br Med J 2:16–20
7. Bassett M, Doran TJ, Halliday JW, Bashir HV, Powell LW (1982) Idiopathic haemochromatosis: demonstration of homozygous-heterozygous mating by HLA typing of families. Hum Genet 60:352–356

8. Bassett ML, Halliday JW, Powell LW (1984) Genetic hemochromatosis. Semin Liver Dis 4:217–227

9. Bassett ML, Halliday JW, Powell LW (1986) Value of hepatic iron measurements in early hemochromatosis and determination of the critical iron level associated with fibrosis. Hepatology 6:24–29

10. Bomford A, Williams R (1976) Long-term results of venesection therapy in idiopathic haemochromatosis. Q J Med 45:611–623

11. Bonkowsky HL, Healey JF, Sinclair PR, Sinclair JF, Pomeroy, JS (1981) Iron and the liver: acute and long-term effects of iron loading on hepatic haem metabolism. Biochem J 196:57–64

12. Borwein ST, Ghent CN, Flanagan PR, Chamberlain MJ, Valberg LS (1983) Genetic and phenotypic expression of hemochromatosis in Canadians. Clin Invest Med 6:171–179

13. Britton RS, Bacon BR, Recknagel RO (1987) Lipid peroxidation and associated hepatic organelle dysfunction in iron overload. Chem Phys Lipids 45:207–239

14. Britton RS, Ferrali M, Magiera CJ, Recknagel RO, Bacon BR (1990a) Increased prooxidant action of hepatic cytosolic low-molecular-weight iron in experimental iron overload. Hepatology 11:1038–1043

15. Britton RS, O'Neill R, Bacon BR (1990b) Hepatic mitochondrial malondialdehyde metabolism in rats with chronic iron overload. Hepatology 11:93–97

16. Britton RS, O'Neill R, Bacon BR (1991) Chronic dietary iron overload in rats results in impaired calcium sequestration by hepatic mitochondria and microsomes. Gastroenterology 101:806–811

17. Cartwright GE, Edwards CQ, Kravitz K, Skolnick M, Amos DB, Johnson A, Buskjaer L (1979) Hereditary hemochromatosis. Phenotypic expression of the disease. N Engl J Med 301:175–179

18. Chojkier M, Houglum K, Solis-Herruzo J, Brenner DA (1989) Stimulation of collagen gene expression by ascorbic acid in cultured human fibroblasts. A role for lipid peroxidation? J Biol Chem 264:16957–16962

19. Cohen A, Witzleben C, Schwartz E (1984) Treatment of iron overload. Semin Liver Dis 4:228–238

20. Dadone MM, Kushner JP, Edwards CQ, Bishop DT, Skolnik MH (1982) Hereditary hemochromatosis: analysis of laboratory expression of the disease by genotype in 18 pedigrees. Am J Clin Pathol 78:196–207

21. de Matteis F, Sparks RG (1973) Iron-dependent loss of liver cytochrome P-450 haem in vivo and in vitro. FEBS Lett 29:141–144

22. Dillard CJ, Tappel AL (1979) Volatile hydrocarbon and carbonyl products of lipid peroxidation: a comparison of pentane, ethane, hexanal and acetone as in vivo indices. Lipids 14:989–995

23. Dillard CJ, Downey JE, Tappel AL (1984) Effects of antioxidants on lipid peroxidation in iron-loaded rats. Lipids 19:127–133

24. Edwards CQ, Griffin LM, Goldgar D, Drummond C, Skolnick MH, Kushner JP (1988) Prevalence of hemochromatosis among 11,065 presumably healthy blood donors. N Engl J Med 318:1355–1362

25. Golberg L, Martin LE, Batchelor A (1962) Biochemical changes in the tissues of animals injected with iron. 3. Lipid peroxidation. Biochem J 83:291–298

26. Gordeuk VR, Bacon BR, Brittenham GM (1987) Iron overload: causes and consequences. Annu Rev Nutr 7:485–508

27. Groskreutz JL, Bronk SF, Gores GJ (1990) Oxidative injury of hepatocytes: mechanisms of lethal injury and protection by ruthenium red. Gastroenterology 98:A590

28. Holland HK, Spivak JL (1989) Hemochromatosis. Med Clin North Am 73:831–845

29. Hruszkewycz AM (1988) Evidence of mitochondrial DNA damage by lipid peroxidation. Biochem Biophys Res Commun 153:191–197

30. Hultcrantz R, Ahlberg J, Glaumann H (1984a) Isolation of two lysosomal populations from iron-overloaded rat liver with different iron concentration and proteolytic activity. Virchows Arch [Cell Pathol] 47:55–65

31. Hultcrantz R, Ericsson JLE, Hirth T (1984b) Levels of malondialdehyde production in rat liver following loading and unloading of iron. Virchows Arch [Cell Pathol] 45:139–146

32. Isaacson C, Seftel HL, Keeley KJ, Bothwell TH (1961) Siderosis in the Bantu. The relationship between iron overload and cirrhosis. J Lab Clin Med 58:845–853

33. LeSage GD, Kost LJ, Barham SS, LaRusso NF (1986) Biliary excretion of iron from hepatocyte lysosomes in the rat: a major excretory pathway in experimental iron overload. J Clin Invest 77:90–97

34. Lindmark B, Eriksson S (1985) Regional differences in the idiopathic hemochromatosis gene frequency in Sweden. Acta Med Scand 218:299–304

35. Louw M, Neethling AC, Percy VA, Carstens M, Shanley BC (1977) Effects of hexachlorobenzene feeding and iron overload on enzymes of haem biosynthesis and cytochrome P-450 in rat liver. Clin Sci 53:111–115

36. Masini A, Ceccarelli-Stanzani D, Trenti T, Ventura E (1984a) Transmembrane potential of liver mitochondria from hexachlorobenzene- and iron-treated rats. Biochim Biophys Acta 802:253–258

37. Masini A, Trenti T, Ventura E, Ceccarelli-Stanzani D, Muscatello U (1984b) Functional efficiency of mitochondrial membrane of rats with hepatic chronic iron overload. Biochem Biophys Res Commun 124:462–469

38. McLaren GD, Muir WA, Kellermeyer RW (1983) Iron overload disorders: natural history, pathogenesis, diagnosis and therapy. CRC Crit Rev Clin Lab Sci 19:205–266

39. Meyer TE, Ballot D, Bothwell TH, Green A, Derman DP, Baynes RD, Jenkins T, Jooste PL, du Toit ED, Jacobs P (1987) The HLA linked iron loading gene in an Afrikaner population. J Med Genet 24:348–356

40. Myers BM, Prendergast FG, LaRusso NF (1988) Experimental iron overload increases the pH of hepatic lysosomes. Hepatology 8:1240

41. Nichols GM, Bacon BR (1989) Hereditary hemochromatosis: pathogenesis and clinical features of a common disease. Am J Gastroenterol 84:851–862

42. Niederau C, Fischer R, Sonnenberg A, Stremmel W, Trampisch HJ, Strohmeyer G (1985) Survival and causes of death in cirrhotic and noncirrhotic patients with primary hemochromatosis. N Engl J Med 313:1256–1262

43. Olsson KS, Ritter B, Rosen U, Heedman PA, Staugard F (1983) Prevalence of iron overload in central Sweden. Acta Med Scand 213:145–150

44. Olsson KS, Eriksson K, Ritter B, Heedman PA (1984) Screening for iron overload using transferrin saturation. Acta Med Scand 215:105–112

45. Park CH, Bacon BR, Brittenham GM, Tavill AS (1987) Pathology of dietary carbonyl iron overload in rats. Lab Invest 57:555–563

46. Peters TJ, O'Connell MJ, Ward RJ (1985) Role of free-radical mediated lipid peroxidation in the pathogenesis of hepatic damage by lysosomal disruption. In: Poli G, Cheeseman KH, Dianzani MU, Slater TF (eds) Free radicals in liver injury. IRL, Oxford, pp 107–115

47. Powell LW, Halliday JW (1980) Idiopathic haemochromatosis. In: Jacobs A, Worwood M (eds) Iron in biochemistry and medicine, vol 2. Academic, New York, pp 461–498

48. Risdon RA, Barry M, Flynn DM (1975) Transfusional iron overload: the relationship between tissue iron concentration and hepatic fibrosis in thalassemia. J Pathol 116:83–95

49. Selden C, Seymour CA, Peters TJ (1980) Activities of some free-radical scavenging enzymes and glutathione concentrations in human and rat liver and their relationship to the pathogenesis of tissue damage in iron overload. Clin Sci 58:211–219

50. Seymour CA, Peters TJ (1978) Organelle pathology in primary and secondary haemochromatosis with special reference to lysosomal changes. Br J Haematol 40:239–253

51. Sharma BK, Bacon BR, Britton RS, Park CH, Magiera CJ, O'Neill R, Dalton N, Smanik P, Speroff T (1990) Prevention of hepatocyte injury and lipid peroxidation by iron chelators and α-tocopherol in isolated iron-loaded rat hepatocytes. Hepatology 12:31–39

52. Shires TK (1982) Iron-induced DNA damage and synthesis in isolated rat liver nuclei. Biochem J 205:321–329
53. Simon M, Alexandre, J-L, Fauchet R, Genetet B, Bourel M (1980) The genetics of hemochromatosis. Prog Med Genet 4:135–168
54. Tanner AR, Desai S, Lu W, Wright R (1985) Screening for haemochromatosis in the UK: preliminary results. Gut 26:1139–1140A
55. Tappel AL (1980) Vitamin E and selenium protection from in vivo lipid peroxidation. Ann NY Acad Sci 355:18–31
56. Tavill AS, Bacon BR (1990) Hemochromatosis: iron metabolism and the iron overload syndromes. In: Zakim D, Boyer TD (eds) Hepatology: a textbook of liver disease. Saunders, Philadelphia, pp 1273–1299
57. Thomas CE, Reed DJ (1988) Effect of extracellular Ca^{++} omission on isolated hepatocytes. II. Loss of mitochondrial membrane potential and protection by inhibitors of uniport Ca^{++} transduction. J Pharmacol Exp Ther 245:501–507

Structural Principles of Flavonoid Antioxidants

W. Bors, W. Heller, C. Michel, and M. Saran

Introduction

Flavonoids are the most ubiquitous and structurally evolved class of plant phenolic compounds. Based on a few principal structures (see Fig. 1), multitudinous hydroxylation, methoxylation, and glycosylation patterns exist. At present more than 4000 individual substances are known [31]. Plants contain both glycosylated compounds and aglycones [74]. As can be expected from the structural diversity, a variety of biological functions have been attributed to flavonoids: photo reception, light screening, visual attraction, feeding repellance, phytoalexin function, etc. [30]. Antioxidative properties were first suggested when it was discovered that the flavonoid content of food contributes to an extended shelf life and retards spoilage [43]. Owing to the rapid degradation of flavonoids in the digestive tract [28, 43], pharmacological effects in mammals are limited [32, 50]. Only few flavonoid derivatives have so far been shown to be of therapeutic value: rutoside [68], cyanidanol (catechin) [20], silybin [69, 70]; all act predominantly as detoxicants after liver injury.

Isoflavonoids, which exist in a few plant families only [21], have also been shown to possess antioxidative properties, e.g., in fermented soybean ("Tempeh" [51]). Structure-activity relationships were studied for the inhibition of lipid peroxidation [40], with the inclusion of synthetic derivatives [75]. An important physiological function of isoflavonoids in plants is to act as precursors for phytoalexins [61]. They will not be included in the subsequent discussion.

The term "antioxidant" implies that the substance counteracts biologically detrimental processes which are ultimately due to toxic properties of oxygen and some of its reduction intermediates. Radicals like superoxide anion (O_2^-) and hydroxyl ($\cdot OH$), their organic analogs peroxyl ($ROO\cdot$) and alkoxyl ($RO\cdot$), as well as the nonradical species singlet oxygen ($^1O_2$) and peroxides (H_2O_2 and $ROOH$) have all been shown to interact with fundamental biological targets, e.g., membrane lipids, proteins, or nucleic acids [27, 54]. In many cases, transition metal complexes catalyze such processes.

Inhibitory effects are based on two basically different chemical intervention mechanisms [7]:

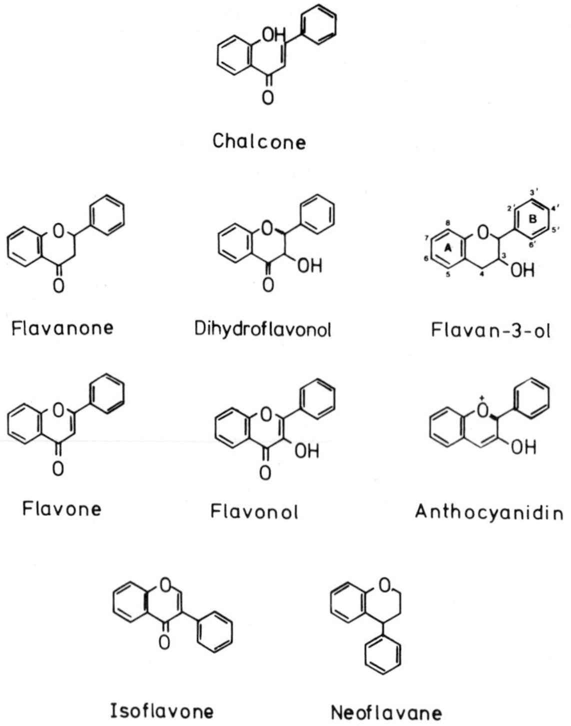

Chalcone

Flavanone   Dihydroflavonol   Flavan-3-ol

Flavone   Flavonol   Anthocyanidin

Isoflavone   Neoflavane

Fig. 1. Major structural classes of flavonoids

1. Prevention of radical formation by:
   - Chelating metal ions
   - Reduction of (hydro)peroxides to innocuous hydroxides
2. Scavenging of radicals by:
   - Formation of less reactive "antioxidant radicals" which disappear by dismutation, recombination, or reduction
   - Catalytic conversion to nonradical products (e.g., the reaction of superoxide dismutase)

Numerous substances have been suggested to act as antioxidants. The most detailed investigations so far were concerned with reactions involving phenolic compounds, ranging from polymer chemistry [47] to biochemistry and food chemistry [18, 59].

Biochemical Studies of Flavonoids in Relation to Their Antioxidative Capacity

Flavonoids as polyphenols may interfere with oxidative processes both by chelating metal ions or by scavenging oxygen radicals. While the first function was an early favorite [19, 49], more recently radical scavenging is considered to be the dominant function of flavonoid antioxidants during the inhibition of lipid peroxidation [2, 55]. Reactions with O_2^- [38, 56], $\cdot OH$ [39], $ROO\cdot$ radicals [64], as well as $^1O_2$ [63, 71] have been investigated. Because in most cases rather unspecific biochemical radical sources were used we decided to study radical reactions of flavonoids, using highly specific photolytic or radiolytic methods to produce oxidizing radicals. Thus far we limited the investigations to flavonoid aglycones.

Radical Reactions and Their Measurement

Generation of Oxidizing Radicals

The basic radiolytic reactions leading to individual radical species in aqueous solutions have been reviewed before [9]. Briefly, the following radicals and respective irradiation conditions were employed (listed in declining order of the electrophilicity of the radicals):

– $\cdot OH$ radical: N_2O-saturated solution
– $\cdot N_3$ radical: N_2O-saturated solution containing $10\,mM$ sodium azide
– tert-Butoxyl (t-BuO$\cdot$) radical: N_2-saturated solution containing $1\,mM$ tert-butylhydroperoxide (t-BuOOH) and $100\,mM$ tert-butanol (t-BuOH) to scavenge $\cdot OH$ radicals
– $SO_3^{\cdot-}$ radical: N_2O-saturated solution containing $10\,mM$ NaHSO$_3$/Na$_2$SO$_3$
– $ROO\cdot$ radicals: solutions saturated with mixtures of N_2O and O_2 and containing $1\,mM$ aliphatic substrates (linoleic acid in strongly alkaline medium and 2-propanol at near neutral pH)
– O_2^- radical: O_2-saturated solution containing $10\,mM$ sodium formate

Alternatively, peroxyl radical was generated by the reaction of $\cdot N_3$ radical with linoleic acid hydroperoxide in strongly alkaline solution [23]. Alkoxyl radicals [t-BuO$\cdot$ and linoleic acid alkoxyl radical and conversion products (LO$\cdot$)] were preferentially produced by UV photolysis of the respective hydroperoxides in the presence of $10\,mM$ t-BuOH to scavenge the simultaneously occurring $\cdot OH$ radical [10]. Azide radical is a highly reactive species, similar to $\cdot OH$ radical. Yet it offers several advantages over the $\cdot OH$ radical, especially in reactions with phenolic substances [3]:

– It is less electrophilic and thus more discriminating in its reactivity than $\cdot OH$.
– It reacts only by electron abstraction from phenolate (or other anionic species).
– It does not add to aromatic structures (as does $\cdot OH$ with the intermediary formation of highly unstable hydroxycyclohexadienyl radicals [1]).

Table 1. Scavenging rate constants of flavonoids for oxidizing radicals

Substance (trivial name)	Rate constant ($\times 10^8\,M^{-1}s^{-1}$)		
	$\cdot OH^a$	$\cdot N_3^b$	$t\text{-BuO}\cdot{}^c$
Flavanols			
1. (+)-Catechin (2R, 3S)	66	50	1.35
2. (−)-Epicatechin (2R, 3R)	64	51	−
Flavanones			
3. Naringenin	210	52	2.65
4. Dihydrofisetin	67	56	−
5. Eriodictyol	117	47	0.8
6. Dihydrokaempferol	58	89	0.95
7. Dihydroquercetin	103	43	1.0
8. Hesperetin	−	58	0.7
Flavylium Salts (Anthocyanidins)			
9. Pelargonidine chloride	45	62	−
10. Cyanidine chloride	−	33	−
Flavones			
11. Apigenin	135	48	3.0
12. Luteolin	130	41	5.7
13. Acacetin	−	28	1.3
Flavonols			
14. Fisetin	−	52	−
15. Kaempferol	141	88	6.0
16. Quercetin	51	66	6.6^d
17. Morin	−	73	−
18. Kaempferid	−	65	−

[a] Determined by pulse radiolysis from the buildup of the aroxyl radical absorption at pH 11.5 ([23], revision and addition of values)
[b] Determined by the same method at pH 11.5 ([8], revision of values)
[c] Data obtained in the "crocin assay" in near neutral solution (pH 6.4–6.8) [10] and recalculated using the reference value for crocin of $3 \times 10^9\,M^{-1}s^{-1}$ [22]
[d] A value of $25 \times 10^8\,M^{-1}s^{-1}$ was determined by pulse radiolysis from the buildup of the aroxyl radical absorption [22]

Scavenging Rate Constants of Flavonoids

Owing to the poor solubility of most flavonoid aglycones at neutral pH, most studies were performed in strongly alkaline solution. Azide and hydroxyl radicals were the preferred oxidizing species employed in the pulse-radiolytic experiments (Table 1). Table 1 also includes rate constants with t-BuO$\cdot$. The values were obtained via the "crocin assay" [10], except for quercetin which was also studied by pulse radiolysis [22]. Kaempferol and quercetin, the most prominent flavonoid radical scavengers, were also investigated with a number of other oxidizing radicals (Table 2).

In Table 1 it can be seen that, with decreasing reactivity of the attacking radical species, the rate constants are spread over a wider range of values

Table 2. Rate constants of the flavonols kaempferol and quercetin with different oxidizing radicals

Substance	Rate constant ($\times 10^8 M^{-1}s^{-1}$)			
	$(SCN)_2 \cdot ^{-a}$	$SO_3 \cdot ^{-b}$	$LOO \cdot ^c$	$O_2 \cdot ^{-d}$
Kaempferol	8.5	4.0	0.34, 0.42	0.0055
Quercetin	4.0	2.5	0.18, 0.15	0.0009

[a] Determined by pulse radiolysis from the kinetics of $(SCN)_2 \cdot ^-$ decay, substrate bleaching, and buildup of the aroxyl radical absorption [23].
[b] Determined by pulse radiolysis from the buildup of the aroxyl radical absorption [24].
[c] Value dependent on method of radical generation, with the first one for a mixture of $LOO \cdot$ radical isomers and the second for 13-$LOO \cdot$ radicals specifically [23].
[d] Determined by pulse radiolysis from the kinetic modeling of the O_2^- decay at pH 7.5 – at higher pH no reaction at all was observed [13].

($f \sim 4$ for $\cdot N_3$ and $f \sim 35$ for t-$BuO \cdot$). From the rate constants with $\cdot N_3$ radical presented in Table 1 and the data shown for kaempferol and quercetin in Table 2, it is obvious that this reactivity of the primary radical is not the only principle which governs rate constants. Rather, for each pair of compounds with 4'-hydroxy and 3',4'-dihydroxy substituents (3/5, 6/7, 9/10, 11/12, 15/16 – see Table 3 for the complete substitution patterns), a consistently higher rate constant is observed for flavonoids with only a *para*-hydroxy group in the B-ring. Such a dependency on the extent of B-ring hydroxylation evidently reflects minor differences in dissociation, as $\cdot N_3$ and the other inorganic radicals as well as $LOO \cdot$ (see Table 2) react only with phenolate groups. The complete set of data for the reactions of quercetin and kaempferol with linoleic acid peroxyl radicals [23] furthermore reveals relatively high rate constants, both for the initial scavenging reactions

$$ArOH + LOO \cdot \rightarrow ArO \cdot + LOOH$$

and for the kinetically observed adduct formation of the peroxyl radicals with flavonoid aroxyl radicals:

$$ArO \cdot + LOO \cdot \rightarrow Ar (=O) OOL$$

The scavenging rate constants exceed those of vitamin E, which were obtained, however, in nonprotic solvents [11].

Transient Spectra of Flavonoids

A qualitative but nevertheless quite revealing structure dependency was observed for the transient spectra of different types of flavonoids after

Table 3. Decay kinetics and spectral parameters of flavonoid aroxyl radicals generated by azide radicals at pH 11.5. [8, 14]

Substance	Substitution		Rate constant	ε	λ
	OH	OCH$_3$	($\times 10^6\,M^{-1}\mathrm{s}^{-1}$)	($M^{-1}\mathrm{cm}^{-1}$)	(nm)
1. (+)-Catechin	3, 5, 7, 3′, 4′	–	0.6	10.000	310
3. Naringenin	5, 7, 4′	–	772[a]	5.350	280
4. Dihydrofisetin	3, 7, 3′, 4′	–	[b]	5.400	315
5. Eriodictyol	5, 7, 3′, 4′	–	0.4	8.900	315
				4.500	360
6. Dihydrokaempferol	3, 5, 7, 4′	–	[c]	4.200	270
7. Dihydroquercetin	3, 5, 7, 3′, 4′	–	0.1	7.600	315
				4.300	360
8. Hesperitin	5, 7, 3′	4′	[d]	5.800	270
9. Pelargonidine chloride	3, 5, 7, 4′	–	210	19.500	685
10. Cyanidine chloride	3, 5, 7, 3′, 4′	–	[e]	4.500	315
				1.800	490
11. Apigenin	5, 7, 4′	–	170	8.100	360
12. Luteolin	5, 7, 3′, 4′	–	0.2	8.100	475
13. Acacetin	5, 7	4′	500	6.600	325
14. Fisetin	3, 7, 3′, 4′	–	0.3	4.100	600
15. Kaempferol	3, 5, 7, 4′	–	140	24.000	550
16. Quercetin	3, 5, 7, 3′, 4′	–	3.4	15.600	530
17. Morin	3, 5, 7, 2′, 4′	–	63	17.600	525
18. Kaempferid	3, 5, 7	4′	370	8.000	480

[a] Slow bleaching at 430 nm of apparent second order with $k/\varepsilon = 5.5 \times 10^4$ cm/s.
[b] First-order decay of 1.45 s^{-1}.
[c] Secondary buildup of transient absorption at 390 nm by first-order process with $(1.2 \pm 0.15) \times 10^3\,\mathrm{s}^{-1}$.
[d] First-order decay of $2.8 \times 10^3\,\mathrm{s}^{-1}$.
[e] Pseudo-first-order decay at 315 nm, assuming reaction with parent compound, gives $7.5 \times 10^5\,M^{-1}\mathrm{s}^{-1}$.

reaction with ·N$_3$ radicals in alkaline aqueous solution (Fig. 2, see also Table 3). Basically, six different spectral types could be observed:

1. Flavanols and flavanones with a saturated 2,3-bond, and containing a 3′,4′-dihydroxy (catechol) structure in the B-ring, exhibit mainly an absorption at 310–315 nm due to the *o*-semiquinone radical (the same major absorption peak is seen for cyanidine chloride, in which case formation of a colorless water adduct interrupts the conjugation through the heterocyclic ring [43]) (group *a*).

2. Flavanones lacking the B-ring catechol structure exhibit an aroxyl radical absorption of the A-ring at 270–280 nm (group *b*).

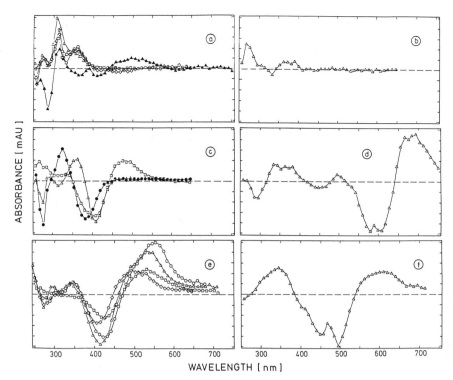

Fig. 2a–f. Transient spectra of flavonoid aroxyl radicals. Dose-normalized spectra, recorded at time of maximal absorption after the pulse; average dose per pulse 9 Gy; attack of azide radicals at pH 11.5: N_2O-saturated solutions, 10 mM NaN$_3$ (substitution pattern given in Table 3). **a** Flavanol, flavanones, and anthocyanidin with B-ring catechol structure (OD scale, ±50 milliabsorbance units full scale. mAUFS). (△) *1*, catechin: concentration 87 μM, observation time 0.04 ms; (□) *4*, dihydrofisetin: 66 μM, 0.04 ms; (◇) *5*, eriodictyol: 48 μM, 40 ms; (○) *7*, dihydroquercetin: 48 μM, 5.3 ms; (▲) *10*, cyanidine chloride: 34 μM, 0.53 ms. **b** *8*, hesperitin (OD scale, ±50 mAUFS): 38 μM, 0.23 ms. **c** Flavones (OD scale, ±100 mAUFS for *11, 12*, ±50 mAUFS for *13*); (△) *11*, apigenin: 45 μM, 0.03 ms; (□) *12*, luteolin: 59 μM, 3.24 ms; (●) *13*, acacetin: 38 μM, 0.18 ms. **d** *9*, pelargonidine chloride (OD scale, ±100 mAUFS): 42 μM, 0.03 ms. **e** Flavonols (= 3-hydroxyflavones; OD scale ±100 mAUFS); (△) *15*, kaempferol: 52 μM, 0.05 ms; (□) *16*, quercetin: 45 μM, 2.03 ms; (○) *17*, morin: 52 μM, 0.03 ms; (◇) *18*, kaempferid: 51 μM 0.05 ms. **f** *14*, fisetin (OD scale ±50 mAUFS): 35 μM, 5.04 ms

3. Flavones with a 2,3-double bond but lacking a 3-hydroxy group reveal an intermediary transient spectrum, dominated by a strong decrease in substrate absorption (group *c*).

4. The anthocyanidin pelargonidine chloride, with the same substitution pattern as the 3-hydroxyflavone (= flavonol) kaempferol, constitutes a class by itself; it does not form a colorless adduct, instead showing the strongest transient absorption and bathochromic shift of all observed aroxyl radicals (group *d*).

5. Flavones, containing both 3- and 5-hydroxy groups, exhibit strong transient absorption and bathochromic shift of the radical absorption, indicative of extensive electron delocalization (group *e*).
6. Fisetin, a 3-hydroxyflavone which lacks a 5-hydroxy group, shows deviations from the transient spectra of the other 3-hydroxyflavones (group *f*).

All flavonoids with a 2,3-double bond and 4-keto group show broad absorption bands extending far into the visible region. Absorption of the *o*-semiquinone structure in the B-ring is completely obscured or – more likely – integrated in the aroxyl radical absorption because of the extended electron delocalization.

Decay Rate Constants of Flavonoid Aroxyl Radicals

A structural dependency is also apparent for the decay rate constants of the flavonoid aroxyl radicals produced with $\cdot N_3$ radicals (Table 3). Generally, the flavonoid aroxyl radicals decay via second-order (dismutation) reactions aside from a few exceptions with first-order decay. The fact that the most stable radicals all contain a catechol structure in the B-ring was the first evidence of radical stability being controlled by structural features.

pH Effects

The above-mentioned results have all been obtained in alkaline solutions, conditions used to improve the solubility of flavonoid aglycones. To study the effectiveness of these compounds as radical scavengers at physiologically relevant pH, we compared a limited number of structural analogs for their reactivity with the 2-propanol peroxyl radical [12]. Several surprising observations resulted from these studies:

1. Table 4 shows that there is a considerable difference in the scavenging rate constants for the individual compounds, with kaempferol far superior to the other substances. A spread of $f \sim 100$ between the highest and lowest value for $(CH_3)_2C(OH)OO\cdot$, as opposed to $f \sim 4$ for $\cdot N_3$ and $f \sim 35$ for $t\text{-BuO}\cdot$ (see Table 1), suggests that this is predominantly an effect of the even lower reactivity of the peroxyl radical [9], which is consequently more discriminating in its attack.
2. Only minor differences are obtained for the decay rate constants of the respective aroxyl radicals, in contrast to the results in alkaline solutions.
3. The transient spectra at both pH conditions are exemplified for kaempferol (Fig. 3) and reveal that the strong visible absorption above 500 nm is absent in neutral solutions after attack by 2-propanol peroxyl radicals. The same effect is observed for the other three flavones and 3-hydroxyflavones (Fig. 4), while neither flavanone nor their radicals absorb in the visible region at all.

Table 4. Rate constants for formation and decay of flavonoid aroxyl radicals: pH effect

Substrate	Rate constant			
	Formation ($\times 10^8\,M^{-1}s^{-1}$)		decay ($\times 10^6\,M^{-1}s^{-1}$)	
	pH 8.5–9.0	pH 11.5	pH 8.5–9.0	pH 11.5
6. Dihydrokaempferol	0.19[a]	33.0	b	c
7. Dihydroquercetin	0.19[a]	24.0	b	0.1
11. Apigenin	0.77	48.0	1.4	170.0
12. Luteolin	1.85	41.0	4.4	0.2
15. Kaempferol	19.5	88.0	4.2	140.0
16. Quercetin	2.1	66.0	3.5	3.4

Experiments at pH 8.5–9.0 with $(CH_3)_2C(OH)OO\cdot$, at pH 11.5 with $\cdot N_3$ as primary radical.
[a] Competition experiments with iron(III) tetrakis-(4-N-methylpyridyl)porphine ($k_{ROO}\cdot$ = $6.8 \times 10^8\,M^{-1}s^{-1}$, ref. [15]) due to low signal to noise ratio.
[b] Very weak signal, decay by apparent first-order reaction with 5×10^4 and $3.4 \times 10^4\,s^{-1}$, respectively.
[c] Secondary buildup of transient absorption at 390 nm by first-order process of $(1.2 \pm 0.15) \times 10^3\,s^{-1}$.

Even by looking at this limited number of structural pairs of flavanones, flavones, and flavonols (4′- vs. 3′,4′- hydroxy groups), it is evident that the stability of the respective transient aroxyl radical is dependent on the presence of the B-ring catechol structure only in strongly alkaline solutions. At lower pH, the decreased extent of dissociation masks this correlation. The previously mentioned effect of B-ring hydroxylation on the rate constant for $\cdot N_3$ attack is not observed for the peroxyl radical at pH 8.5, with the exception of kaempferol and quercetin. More data are needed to resolve this inconsistency.

Electron Spin Resonance

It is apparent from these results that observation of visible and UV transient spectra alone is insufficient to quantify correlations between structural and kinetic features, especially for the highly substituted tetra- to hexa-hydroxylated flavonoids. However, electron spin resonance (ESR) spectroscopy as the only convenient method for such studies has only been employed once [37]. In the meanwhile we have initiated ESR studies as well. Problems such as stability of the aroxyl radicals, multiple sites of attack for different types of oxidizing radicals, and extensive electron delocalization in the conjugated system have to be carefully considered before a reasonable interpretation of the spectra can be achieved.

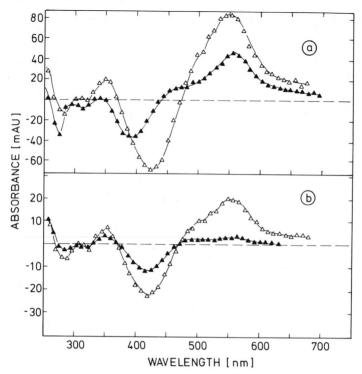

Fig. 3a,b. Transient spectra of the kaempferol aroxyl radical. Dose-normalized spectra, recorded at time of maximal absorption after the pulse. *Solid symbols*, pH 8.0–8.5; *open symbols*, pH 11.5. **a** Azide radicals, N_2O-saturated solutions with $10\,mM$ NaN_3: (▲) pH 8.0, concentration of kaempferol $62\,\mu M$, observation time 0.53 ms, pulse dose 15 Gy; (△) pH 11.5, $52\,\mu M$, 0.035 ms, 8 Gy. **b** Peroxyl radicals, solutions saturated with various $N_2O:O_2$ mixtures: (▲) pH 8.5, concentration of 2-propanol $100\,mM$, of kaempferol $58\,\mu M$, 1.03 ms, 29.4 Gy; (△) pH 11.5, concentration of linoleic acid $1\,mM$, of kaempferol $49\,\mu M$, 0.45 ms, 11.5 Gy

Product Studies

It is a long mechanistic pathway from the initially observable aroxyl radicals to stable products which might be isolated and identified. For example, products proposed to result from the reaction of O_2^- with kaempferol, leading to a fragmentation of the heterocyclic ring, are certainly not derived from O_2^- reactions alone [62]. Firstly, the reaction of O_2^- is rather slow [13] and, secondly, under the conditions used by Takahama [62], metal catalysis cannot be excluded which might result in the adventitious formation of ·OH radicals.

Our attempts to isolate adducts of quercetin or kaempferol aroxyl radicals with linoleic acid peroxyl radicals proved futile so far. Kinetic evidence strongly argues for the existence of such adducts [23]. Yet high-

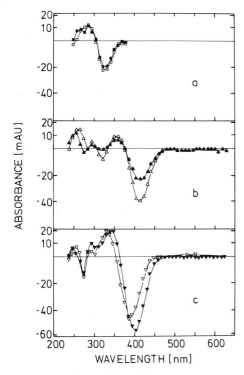

Fig. 4a–c. Comparison of transient spectra of structurally paired flavonoids. Dose-normalized uncorrected spectra in unbuffered aqueous solution, pH 8.5–9.0, containing 10 mM 2-propanol and saturated with various $N_2O:O_2$ mixtures. Observation time at maximum transient absorption after the pulse (*open symbols*, 4'-OH substitution; *solid symbols*, 3',4'-di-OH substitution in the B-ring). **a** *Dihydroflavonols*: (●) *6*, dihydro-kaempferol: substrate concentration 18 μM, concentration of $(CH_3)_2C(OH)OO\cdot$ radical 6.5 μM, pH 9.0, observation time 5 μs; (■) *7*, dihydroquercetin: 18 μM, 5.8 μM, pH 9.0, 5 μs. **b** *Flavonols*: (△) *15*, kaempferol: 22 μM, 17 μM, pH 8.5, 540 μs; (▲) *16*, quercetin: 25 μM, 9.0 μM, pH 8.5, 210 μs. **c** *Flavones*: (▽) *11*, apigenin: 25 μM, 10.3 μM, pH 8.5, 120 μs; (▼) *12*, luteolin: 24 μM, 9.1 μM, pH 8.5, 5 μs

performance liquid chromatography (HPLC) of the reaction mixture resulted in several fractions with properties of hydroperoxides, which were too unstable to last through further purification.

Structural Principles

Results from Radio- and Photolytic Studies

It may be considered fortuitous that our investigations started with alkaline solutions of the flavonoids. As compared to the limited data in neutral or slightly alkaline solutions, the presence of several dissociated

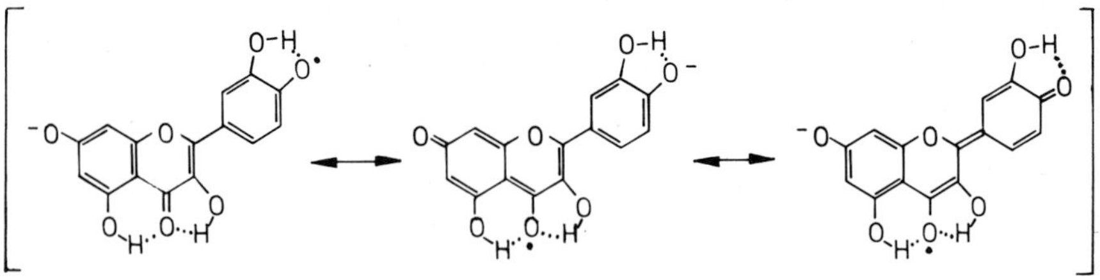

Fig. 5a–c. Mesomeric structures of the quercetin aroxyl radical

phenolic hydroxy groups strongly favors extensive electron delocalization, apparent as:

– Aroxyl radicals which show intense absorption in the visible region
– Decay rate constants, which clearly depend on the number and position of hydroxy groups

From the accumulated data, we proposed [13] that the basic requirements for optimal antioxidative potential, i.e., scavenging efficiency *and* relative stability of the formed flavonoid aroxyl radicals, are:

– Conjugation of the 2,3-double bond and the 4-oxo group
– Hydrogen bonding between this 4-oxo group with neighboring 3- and 5-hydroxy groups
– A 3′,4′-dihydroxy (catechol) structure in the B-ring

The association of stability with extensive electron delocalization of the radical structure is demonstrated in Fig. 5 for three mesomeric structures of the quercetin aroxyl radical.

The data for neutral solutions using the less reactive, but more discriminating, 2-propanol peroxyl radical as oxidizing agent are basically in line with this hypothesis – in particular with respect to the influence of dissociated phenoxy groups. Since dissociation is limited in neutral solutions – based on the accepted dissociation sequence 7-OH $\gg$ 4′-OH > 3-OH [46, 60] – less prominent radical absorption and minimal difference of the decay rates are to be expected. The exceptionally high rate constant of kaempferol with the $(CH_3)_2C(OH)OO\cdot$ radical, however, cannot be fully rationalized with these arguments and requires confirmation by studies using more structural analogs or other oxidizing radicals.

Natural flavonoids, with very few exceptions, all contain the 7-hydroxy group. This is due to the biosynthetic pathway, where three acetate units are condensed to a hydroxycinnamic acid moiety. Eventually this leads to 5,7-dihydroxylated flavonoids or, by reduction of one of the carbonyl function of the intermediary polyketide, to a 5-deoxy structure [4, 33, 72]. This suggests that, in view of these highly evolved compounds [31], retaining the 7-hydroxy group must be of importance for the various biological

Table 5. Structure-activity relationships for the inhibition of enzymes by flavonoids

Enzyme	EC	Number of flavonoids tested	Structural principles for optimal inhibition	Ref.
Oxidoreductases				
Aldose reductase	1.1.1.21	34	Methoxylated	[53]
		44	Glycosides of penta-OH compounds	[65]
Succinoxidase	1.3.99.1	14	3',4'-di-OH	[34]
NADH: coenzyme Q reductase	1.6.99.2	14	3',4'-di-OH + 2,3-ene + 4-oxo	[35]
Lipoxygenase	1.13.11.12	7	2,3-ene + 3-OH (analogy to nordihydroguaiaretic acid)	[73]
Cyclooxygenase	1.14.99.1	22(+15)	o-di-OH in A- or B-ring	[5]
Cytochrome P-450	1.14.99.–	16	hydrophilic	[17]
		15	2,3-ene + several OH	[66]
Transferases				
Catechol-O-methyltransferase	2.1.1.6	19(+5)	3- + 5-OH + 4-oxo + 2,3-ene + 3',4'-di-OH	[58]
Protein kinases	2.7.1.37	7	Differential effect of OH pattern on isoenzymes	[29]
		15	2,3-ene + 7-OH + 3',4'-di-OH	[26]
Phosphorylase kinase	2.7.1.38	14	Quercetin, fisetin are best	[44]
Hydrolases				
Phosphodiesterase	3.1.4.16	17(+12)	Selectivity for cGMP vs. cAMP enzyme	[57]
		13	2,3-ene + 4-oxo + 3-OH + moderate hydroxylation	[6]
		45	Purine structural analogs: quercetin, cyanidin are best	[25]
(+ other enzymes)		15	2,3-ene + 3-OH (flavonol aglycones)	[45]
Lyases				
Glyoxalase I	4.4.1.5	13(+48)	2,3-ene + 3-OH + 4-oxo + 7-OH (no -OCH$_3$)	[42]

functions of flavonoids. Incidentally, 7-deoxyflavonoids are so far known only from the farina of Primulaceae and from Rutaceae [73].

Comparison with Other Structure-Activity Relationships

The diversity of structural features obviously makes flavonoids good candidates for structure-activity relationship (SAR) studies, especially since they have demonstrated, aside from acting as antioxidants, a broad pharmacological action potential [32, 50].

Enzyme Inhibition

Individual flavonoids – mostly quercetin – have been shown to inhibit enzymes belonging to such diverse classes as oxidoreductases, transferases, hydrolases, or lyases. In Table 5, results of SAR studies are compiled which were performed with 11 different enzymes or isoenzyme groups (cytochromes P-450, protein kinases, cyclic nucleotide phosphodiesterases). The most thorough SAR within each enzyme class [6, 35, 42, 58] came up with the same structural features that we found for the most effective radical-scavenging flavonoid antioxidants [13]. Exceptions were the inhibitors for aldose reductase, where hydrophobicity seems to be the dominant structural feature which increased the inhibitory activity [53, 65]. In the case of cytochrome P-450, hydrophobic flavonoid compounds were activators of enzyme activity [17]. The highly selective effects on individual protein kinases [29] or cAMP- vs. cGMP-phosphodiesterases [45, 57] are of considerable interest for a potential regulatory function of flavonoids. The technically most advanced SAR studies included a quantum-mechanical comparison of flavonoids with the cAMP structure [25] and a computer-automated structure evaluation ("CASE" [42]) of flavonoids as ene-diol substrate analogs for glyoxalase I.

Autoxidation

Pardini and collaborators studied the autoxidative and electrochemical behavior of flavonoids [36] and related it to the inhibition of mitochondrial respiration and formation of oxygen radicals [34, 35]. Again, similar structural criteria as for antioxidant behavior were noted, i.e., the presence of a 2,3–double bond, a 4-oxo, and/or a 3-hydroxy group [36]. This independent confirmation of our SAR for antioxidative behavior points to similar mechanisms, i.e., the intermediary formation of aroxyl radicals in univalent redox steps during both radical scavenging and autoxidation and electrochemical reactions.

Mutagenicity

Structure-activity relationship studies have also been conducted to determine the promutagenic behavior of flavonoids [16, 48, 52]. Yet, despite all attempts, aside from the finding that quercetin is the strongest mutagen among flavonoids in the bacterial "Ames test" [52], no clear-cut structural principles could be discerned.

Conclusions

Detailed kinetic studies on the scavenging efficiencies of a number of flavonoids toward individual oxidizing radicals resulted in substantiating their capacities to act as antioxidants. They furthermore pointed to distinct SARs, which are also valid for chromane model compounds [14].

The fact that basically the same structural criteria have been noted for inhibition of enzymes belonging to different classes may indicate that in each case aroxyl radical intermediates are somehow involved. Only for oxidoreductases and especially for the flavonoid-sensitive lipoxygenase [67] and cyclooxygenase [41] has intermediary radical formation during the catalytic cycle been reported. Aroxyl radical formation from flavonoids could thus only arise during adventitious autoxidation reactions [36], if one extends this argument to the other enzyme classes.

A structural analogy to natural substrates has been the preferred hypothesis in these cases [25, 42, 58]. As an alternative to radical formation, one may argue that the common denominator for the various activities of flavonoids requiring basically the same structures is the presence of these highly conjugated π-electron systems, combined with a sufficient extent of hydroxylation. Kinetic spectroscopy on the formation and decay of radical species thus complements information on the state of electron delocalization and/or resonance structure in a given molecule.

In view of the many gaps still existing in our knowledge on the detailed mechanisms for the inhibitory activities of flavonoids, the versatility of the pulse radiolysis method combined with ESR studies, should enable us to further close these gaps.

References

1. Adams GE, Michael BD (1976) Pulse radiolysis of benzoquinone and hydroquinone. Semiquinone formation by water elimination from trihydroxycyclohexadienyl radicals. Trans Faraday Soc 63:1171–1180
2. Affany A, Salvayre R, Douste-Blazy L (1987) Comparison of the protective effect of various flavonoids against lipid peroxidation of erythrocyte membranes (induced by cumene hydroperoxide). Fundam Clin Pathol 1:451–457
3. Alfassi ZB, Schuler RH (1985) Reaction of azide radicals with aromatic compounds. Azide as a selective oxidant. J Phys Chem 89:3359–3363
4. Ayabe S-I, Udagawa A, Furuya T (1988) NAD(P)H-dependent 6'-deoxychalcone synthase activity in *Glycyrrhiza echinata* cells induced by yeast extract. Arch Biochem Biophys 261:458–462

5. Baumann J, Wurm G, von Bruchhausen F (1980) Hemmung der Prostaglandin-synthetase durch Flavonoide und Phenolderivate im Vergleich mit deren O_2^- Radikalfängereigenschaften. Arch Pharm 313:330–337

6. Beretz A, Cazenave JP, Anton R (1982) Inhibition of aggregation and secretion of human platelets by quercetin and other flavonoids: structure-activity relationships. Agents Actions 12:382–387

7. Bors W (1986) Bedeutung und Wirkungsweise von Antioxidantien. In: Elstner EF, Bors W, Wilmanns W (eds) Reaktive Sauerstoffspezies in der Medizin. Springer, Berlin Heidelberg New York, pp 161–183

8. Bors W, Saran M (1987) Radical scavenging by flavonoid antioxidants. Free Radic Res Commun 2:289–294

9. Bors W, Saran M, Michel C, Tait D (1984a) Formation and reactivities of oxygen free radicals. In: Breccia A, Greenstock GL, Tamba M (eds) Advances on oxygen radicals and radioprotectors. Lo Scarabeo Bologna, pp 13–27

10. Bors W, Michel C, Saran M (1984b) Inhibition of the bleaching of the carotenoid crocin. A rapid test for quantifying antioxidant activity. Biochim Biophys Acta 796:312–319

11. Bors W, Erben-Russ M, Saran M (1987) Fatty acid peroxyl radicals: their generation and reactivities. Bioelectrochem Bioenerg 18:37–49

12. Bors W, Heller W, Michel C, Saran M (1989) Reactions of flavonoid antioxidants with peroxyl radicals. 3rd. Int Conference on anticarcinogens and radiation protection. Dubrovnik, Oct. 15–21, 1989

13. Bors W, Heller W, Michel C, Saran M (1990a) Flavonoids as antioxidants: determination of radical-scavenging efficiencies. In: Packer L, Glazer AN (eds) Methods in enzymology, vol 186. Academic, New York, pp 343–355

14. Bors W, Heller W, Michel C, Saran M (1990b) Radical chemistry of flavonoid antioxidants. In: Emerit I, Packer L, Auclair C (eds) Antioxidants in therapy and preventive medicine. Plenum, New York, pp 165–170

15. Bors W, Czapski G, Saran M (1991) An expanded function for superoxide dismutase. Free Radic Res Commun 12–13:411–417

16. Brown JP (1980) A review of the genetic effects of naturally occurring flavonoids, anthraquinones and related compounds. Mutat Res 75:243–277

17. Buening MK, Chang RL, Huang MT, Fortner JG, Wood AW, Conney AH (1981) Activation and inhibition of benzo[a]pyrene and aflatoxin B1 metabolism in human liver microsomes by naturally occurring flavonoids. Cancer Res 41:67–72

18. Burlakova EB (1975) Bioantioxidants and synthetic inhibitors of radical processes. Russ Chem Rev 44:871–880

19. Clemetson CAB, Andersen L (1966) Plant polyphenols as antioxidants for ascorbic acid. Ann NY Acad Sci 136:339–378

20. Conn HO (ed) (1984) (+)-Cyanidanol-3 in diseases of the liver. Proceedings of the International Workshop. Grune & Stratton, San Francisco

21. Dewick PM (1988) Isoflavonoids. In: Harborne JB (ed) The flavonoids. Advances in research since 1980. Chapman & Hall, London, pp 125–209

22. Erben-Russ M, Michel C, Bors W, Saran M (1987a) Absolute rate constants of alkoxyl radical reactions in aqueous solutions. J Phys Chem 91:2362–2365

23. Erben-Russ M, Bors W, Saran M (1987b) Reactions of linoleic acid peroxyl radicals with phenolic antioxidants: a pulse radiolysis study. Int J Radiat Biol 52:393–412

24. Erben-Russ M, Michel C, Bors W, Saran M (1987c) Determination of sulfite radical reaction rate constants by means of competition kinetics. Radiat Environ Biophys 26:289–294

25. Ferrell JE, Chang Sing PDG, Loew G, King R, Mansour JM, Mansour TE (1979) Structure/activity studies of flavonoids as inhibitors of cAMP phosphodiesterase and relationship to quantum chemical indices. Mol Pharmacol 16:556–568

26. Ferriola PC, Cody V, Middleton E (1989) Protein kinase C inhibition by plant flavonoids. Kinetic mechanisms and structure activity relationships. Biochem Pharmacol 38:1617–1624

27. Grisham MB, McCord JM (1986) Chemistry and cytotoxicity of reactive oxygen metabolites. In: Taylor AE, Matalon S, Ward PA (eds) Physiology of oxygen radicals. Am Physiol Soc, Bethesda, pp 1–18
28. Hackett AM (1986) The metabolism of flavonoid compounds in mammals. In: Cody V, Middleton E, Harborne JB (eds) Plant flavonoids in biology and medicine. Liss, New York, pp 177–194
29. Hagiwara M, Inoue S, Tanaka T, Nunoki K, Ito M, Hidaka H (1988) Differential effects of flavonoids as inhibitors of tyrosine protein kinases and serine/threonine protein kinases. Biochem Pharmacol 37:2987–2992
30. Harborne JB (1986) Nature, distribution and function of plant flavonoids. In: Cody V, Middleton E, Harborne JB (eds) Plant flavonoids in biology and medicine. Liss, New York, pp 15–24
31. Harborne JB (ed) (1988) Flavonoids: advances in research since 1980. Chapman & Hall, London
32. Havsteen B (1983) Flavonoids, a class of natural products of high pharmacological potency. Biochem Pharmacol 32:1141–1148
33. Heller W, Forkmann G (1988) Biosynthesis. In: Harborne JB (ed) The flavonoids. Advances in research since 1980. Chapman & Hall, London, pp 399–425
34. Hodnick WF, Kung FS, Roettger WJ, Bohmont CW, Pardini RS (1986) Inhibition of mitochondrial respiration and production of toxic oxygen radicals by flavonoids. Biochem Pharmacol 35:2345–2357
35. Hodnick WF, Bohmont CW, Capps C, Pardini RS (1987) Inhibition of the mitochondrial NADH-oxidase (NADH-coenzyme Q oxidoreductase) enzyme system by flavonoids: a structure activity study. Biochem Pharmacol 36:2873–2874
36. Hodnick WF, Milosavljevic EB, Nelson JH, Pardini RS (1988a) Electrochemistry of flavonoids. Relationships between redox potentials, inhibition of mitochondrial respiration, and production of oxygen radicals by flavonoids. Biochem Pharmacol 37:2607–2611
37. Hodnick WF, Kalyanaraman B, Pritsos CA, Pardini RS (1988b) The production of hydroxyl and semiquinone free radicals during the autoxidation of redox active flavonoids. In: Simic MG, Taylor KA, Ward JF, von Sonntag C (eds) Oxygen radicals in biology and medicine, (Basic life sciences, vol 49) Plenum, New York, pp 149–152
38. Huguet AI, Manez S, Alcaraz MJ (1990) Superoxide scavenging properties of flavonoids in a non-enzymic system. Z Naturforsch 45c:19–24
39. Husain SR, Cillard J, Cillard P (1987) Hydroxyl radical scavenging activity of flavonoids. Phytochemistry 26:2489–2491
40. Jha, HC, von Recklinghausen G, Zilliken F (1985) Inhibition of in vitro microsomal lipid peroxidation by isoflavonoids. Biochem Pharmacol 34:1367–1369
41. Kalyanaraman B, Sivarajah K (1984) The ESR study of free radicals formed during the arachidonic acid cascade and cooxidation of xenobiotics by prostaglandin synthase. In: Pryor WA (ed) Free radicals in biology, vol 6. Academic, New York, pp 149–198
42. Klopman G, Dimayuga ML (1988) Computer-automated structure evaluation of flavonoids and other structurally related compounds as glyoxalase I enzyme inhibitors. Mol Pharmacol 34:218–222
43. Kuehnau J (1976) The flavonoids. A class of semi-essential food components: their role in human nutrition. World Rev Nutr Diet 24:117–191
44. Kyriakidis SM, Sotiroudis TG, Evangelopoulos AE (1986) Interaction of flavonoids with rabbit muscle phosphorylase kinase. Biochim Biophys Acta 871:121–129
45. Landolfi R, Mower RL, Steiner M (1984) Modification of platelet function and arachidonic acid metabolism by bioflavonoids. Biochem Pharmacol 33:1525–1530
46. Mabry TJ, Markham KR, Thomas MR (eds) (1970) The systematic identification of flavonoids vol 2. Springer, Berlin Heidelberg New York
47. Mahoney LR (1969) Antioxidantien. Angew Chem 81:555–563
48. McGregor JT (1986) Mutagenic and carcinogenic effects of flavonoids. In: Cody V, Middleton E, Harborne JB (eds) Plant flavonoids in biology and medicine. Liss, New York, pp 411–424

49. Mehta AC, Seshadri TR (1959) Flavonoids as antioxidants. J Sci Ind Res 18B: 24–28
50. Middleton E (1984) The flavonoids. Trends Pharmacol Sci 5:335–338
51. Murakami H, Asakawa T, Terao J, Matsushita S (1984) Antioxidative stability of tempeh and liberation of isoflavones by fermentation. Agric Biol Chem 48:2971–2975
52. Nagao M, Morita N, Yahagi T, Shimizu M, Kuroyanagi M, Fukuoka M, Yoshihira K, Natori S, Fujino T, Sugimura T (1981) Mutagenicities of 61 flavonoids and 11 related compounds. Environ Mutagen 3:401–419
53. Okuda J, Miwa I, Inagaki K, Horie T, Nakayama M (1982) Inhibition of aldose reductases from rat and bovine lenses by flavonoids. Biochem Pharmacol 31: 3807–3822
54. Pryor WA (1986) Oxy-radicals and related species: their formation, lifetimes, and reactions. Annu Rev Physiol 48:657–667
55. Ratty AK, Das NP (1988) Effects on flavonoids on nonenzymatic lipid peroxidation: structure-activity relationship. Biochem Med Metab Biol 39:69–79
56. Robak J, Gryglewski RJ (1988) Flavonoids are scavengers of superoxide anions. Biochem Pharmacol 37:837–841
57. Ruckstuhl M, Landry Y (1981) Inhibition of lung cyclic AMP- and cyclic GMP-phosphodiesterases by flavonoids and other chromone-like compounds. Biochem Pharmacol 30:697–702
58. Schwabe KP, Flohé L (1972) Catechol-O-methyltransferase. III. Beziehung zwischen der Struktur von Flavonoiden und deren Eignung als Inhibitoren der Catechol-O-Methyltransferase. Z Physiol Chem 353:476–482
59. Sherwin ER (1972) Antioxidants for food fats and oils. J Am Oil Chem Soc 49: 468–472
60. Slabbert NP (1977) Ionisation of some flavanols and dihydroflavanols. Tetrahedron 33:821–824
61. Smith DA, Bank SW (1986) Biosynthesis elicitation and biological activity of iso-flavonoid phytoalexins. Phytochemistry 25:979–995
62. Takahama U (1987) Oxidation products of kaempferol by superoxide anion radical. Plant Cell Physiol 28:953–957
63. Takahama U, Youngman RJ, Elstner EF (1984) Transformation of quercetin by singlet oxygen generated by a photosensitized reaction. Photobiochem Photobiophys 7:175–181
64. Torel J, Cillard J, Cillard P (1986) Antioxidant activity of flavonoids and reactivity with peroxy radicals. Phytochemistry 25:383–385
65. Varma SD, Kinoshita JH (1976) Inhibition of lens aldose reductase by flavonoids – their possible role in the prevention of diabetic cataracts. Biochem Pharmacol 25:2505–2513
66. Vernet A, Siess MH (1986) Comparison of the effects of various flavonoids on ethoxycoumarin deethylase activity of rat intestinal and hepatic microsomes. Food Chem Toxicol 24:857–861
67. Vliegenthart JFG (1979) Enzymic and non enzymic oxidation of polyunsaturated fatty acids. Chem Ind 241–251
68. Voelter W, Jung G (eds) (1978) O-(β-Hydroxyethyl)-rutoside – experimentelle und klinische Ergebnisse. Springer, Berlin Heidelberg New York
69. Vogel G, Trost W, Braatz R, Odenthal KP, Brüsewitz G, Antweiler H, Seeger R, Ulbrich M (1975) Untersuchungen zu Pharmakodynamik, Angriffspunkt und Wirkungsmechanismus von Silymarin, dem anti-hepatotoxischen Prinzip aus *Silybum marianum* (L.) Gaertn. 1. Mitt.: Akute Toxikologie bzw. Verträglichkeit, allgemeine und spezielle (Leber-)Pharmakologie. 2. Mitt.: Besondere Untersuchungen zu Angriffspunkt und Wirkungsmechanismus. Arzneim Forsch 25:82–90, and 179–187
70. Wagner H (1986) Antihepatotoxic flavonoids. In: Cody V, Middleton E, Harborne JB (eds) Plant flavonoids in biology and medicine. Liss, New York, pp 545–558
71. Wagner GR, Youngman RJ, Elstner EF (1988) Inhibition of chloroplast photo-oxidation by flavonoids and mechanisms of the antioxidative action. J Photochem Photobiol B 1:451–460

72. Welle R, Grisebach H (1988) Isolation of a novel NADPH-dependent reductase which coacts with chalcone synthase in the biosynthesis of 6'-deoxychalcone. FEBS Lett 236:221–225
73. Wheeler EL, Berry DL (1986) In vitro inhibition of mouse epidermal cell lipoxygenase by flavonoids: structure-activity relationships. Carcinogenesis 7:33–36
74. Wollenweber E, Dietz VH (1981) Occurrence and distribution of free flavonoid aglycones in plants. Phytochemistry 20:896–932
75. Zilliken FW (1981) Isoflavones and related compounds, methods of preparing and using and antioxidant compositions containing same. United States Patent No 4, 264, 509

Role of Free Radical Reactions in Experimental Hyperlipidemia in the Pathomechanism of Fatty Liver

A. Blázovics, E. Fehér, and J. Fehér

Introduction

The oxygen free radicals, which play an important role in the living organism, are generated from oxygen by excitation or reduction. The primary free radicals generated are univalent and divalent products of oxygen, O_2^-, and H_2O_2. The true free radicals O_2^-, $\cdot OH$, and reactive products $^1O_2$ or H_2O_2 may be formed by exogenous interactions, but also during physiological processes in all parts of the cells in the living organism [9, 15, 22, 28, 41, 42].

Exposure of cell membranes to oxygen radicals stimulates the process of lipid peroxidation. Iron and copper catalyze these chain reactions and thus accelerate the lipid peroxidation [22, 37].

Physiological free radical reactions are under control [33]. If the protective mechanisms of the organism are not sufficiently active as a result of the attack by free radicals, tissue damage develops. Pathological free radical reactions are uncontrolled reactions in the cell. [11, 20].

Free radical reactions have been found to be involved in the pathogenesis of certain liver, kidney, lung, brain, and age-related vascular diseases. Free radicals can also be produced in the cell as a result of xenobiotic metabolism [10, 15, 23, 24].

Much evidence has been presented that the lipid peroxides accumulated in the blood and in the arterial wall cause the inhibition of natural defense mechanisms and stimulate the development of atherosclerosis [15, 24].

Free radicals may have a part in the pathogenesis of fatty liver, increasing metabolic alterations. The damage caused by the free radicals can be prevented by scavenger molecules, which are natural or synthetic antioxidants; thus complementary treatment with drugs of this group can be suggested in diseases where free radicals play a leading role.

In our previous experiments we reported the antioxidant effects of the CH 402 [48] (Na-2,2-dimethyl-1,2-dihydroquinoline-4-yl methane sulfonate) and MTDQ-DA [47] (6,6'-methylene-bis 2,2-dimethyl-4-methane sulfonic acid: Na-1,2-dihydroquinoline) dihydroquinoline-type chemicals on different tissues and subcellular fractions in vivo and in vitro. We found an inhibitory effect of CH 402 and MTDQ-DA derivatives and enzymatic and non-enzymatic lipid peroxidation in different systems in vitro in the liver and in the brain as well as their membrane protective effects [3–5, 12, 46].

The use of MTDQ-DA in cholesterol-induced experimental athero-sclerosis (rabbit) significantly reduced serum cholesterol and triglyceride levels. At the same time, it increased the high-density lipoprotein (HDL) fraction and decreased serum malondialdehyde and serum and granulocyte β-glucuronidase levels (these are used as indicators of peroxidative damage) [13].

Silibinin is a well-known free radical scavenger and antioxidant [2, 6, 15, 17, 49]. In animal experiments it has proved to be effective in carbon tetrachloride, galactosamine, polycyclic aromatic hydrocarbon, and ethanol induced liver damage, as well as in *Amanita phalloides* poisoning. This drug is widely used in the therapy of liver diseases of different etiology. It also has noncompetitive lipoxygenase and prostaglandin synthetase inhibitor activity which contributes to the drug's antiinflammatory effect in chronic hepatitis [7, 8, 16, 18, 34].

The aim of the present work is to verify the damaging effect of free radical reactions and lipid peroxidation in experimental hyperlipidemia in rats, and summarize knowledge on the mechanism of two free radical scavengers developed in Hungary, the CH 402 and MTDQ-DA dihydro-quinoline-type derivatives, and silibinin (Madaus, Cologne), a natural flavonolignane-type derivate, as a control.

The free radical scavenger effects of these antioxidants were studied in different free radical generating in vitro systems by means of spectro-photometry and measurement of chemiluminescence. In these experiments we studied the effects of CH 402, MTDQ-DA, and silibinin in vivo and in vitro in normo- and hyperlipidemic rat liver microsomes on the microsomal monooxygenase enzyme system, on induced lipid peroxidation, and on changes in lipid parameters of microsomal membranes in liver and blood; we aimed to elucidate the mechanism of action and to assess whether the ultrastructural and morphological changes correlated with the biochemical parameters of the monooxygenase system.

Materials and Methods

Young male Wistar rats weighing about 200 g were used for all experiments. They were fed on a fat-rich diet containing 2% cholesterol, 0.5% cholic acid, and 20% sunflower oil added to the normal "Lati" chow (Hungary) for 8 days. The animals of the first group were kept on normal chow, those of the second group were kept on a normal diet and treated with CH 402 or MTDQ-DA 200 mg/kg per day via a gastric tube or 25 mg/kg per day silibinin i.p. injection for 5 days. The third group was fed on a fat-rich diet for 8 days, and the animals of the fourth group were fed on a fat-rich diet and treated with CH 402, MTDQ-DA, or silibinin.

Histological and Biochemical Methods

Histological Examinations. Liver fragments were fixed in 4% neutral buffered formalin, embedded in paraffin, and 5-µm sections were cut and stained with hematoxylin-eosin. Electron microscopy specimens of the liver were fixed with a fixative containing 2.5% paraformaldehyde and 2.5% glutaraldehyde. The samples were postfixed in 1% osmium tetroxide and embedded in Epon. Ultrathin sections were counterstained with uranyl acetate and lead citrate and examined under a Tesla BS500 electron microscope.

Liver Microsomes. The microsomes were prepared by ultracentrifugation in $0.15\,M$ KCl solution [14].

Spectrophotometry. The enzymatically induced lipid peroxidation was measured in a medium of total vol. 0.5 ml with a protein content of 1 mg/ml. The medium contained $20\,\text{m}M$ sodium phosphate buffer, pH 7.5, $0.15\,\text{m}M$ KCl, $50\,\mu M$ FeCl$_3$, $50\,\mu M$ sodium pyrophosphate glucose-6-phosphate dehydrogenase 0.6 IU, $0.5\,\text{m}M$ NADPH, and $10\,\text{m}M$ glucose-6-phosphate, plus various concentrations of CH 402, MTDQ-DA, or silibinin [4].

Malondialdehyde (MDA) production was monitored by the thiobarbituric acid test of Ottolenghi [40], NADPH cytochrome c reductase, NADH ferricyanide reductase [29], and aminopyrine-N-demethylase [36]; cytochrome P-450 content was also measured [38, 39] by the standard methods.

The lipid parameters in the microsomal membranes were measured by the Merckotest methods. In the sera these parameters were measured using the Centrifichem 500 systems, (total cholesterol and HDL cholesterol with the Baker kit, triglycerides with the *SKI* kit). Serum HDL cholesterol was assayed after precipitations of low-density lipoprotein (LDL) and very low density lipoprotein (VLDL) fractions with Merck precipitating reagent. Serum LDL cholesterol was determined according to Friedewald et al. [21].

The effect of the antioxidants on adrenochrome formation during autooxidation of epinephrine was followed by monitoring the light absorbance of the product at a wavelength of 480 nm. The medium consisted of $1.6 \times 10^{-3}\,M$ epinephrine, $50\,\text{m}M$ sodium carbonate buffer (pH 10.1), and antioxidants at various concentrations [35]. Diene conjugation was measured using the AOAC methods [1].

The protein concentration of the preparation was determined by Lowry et al. [32] using serum albumin as a standard.

Chemiluminescence Measurement. Light emission was measured by the chemiluminescence measurement method with a CLD-1 Medicor-Medilab luminometer (Hungary).

Luminescent light is measured with a sensitive photomultiplier. The electrical signals of the latter are processed by means of an MMT microprocessor system, and the concentration of the investigated substance is indicated digitally. The apparatus ensures the maintenance of a standard temperature of the investigated substance and its homogenization. The apparatus is suitable for the performance of the following four measuring programs:

- Continuous measurement of the light intensity
- Measurement of the peak value of the light intensity in a rapid reaction
- Measurement of the rate in light intensity changes
- Measurement of the integrated value of intensity change during a given time interval [50]

The measurements were carried out in a double-part cuvette [4, 17], the contents of each part being mixed by centrifugation started at the same time as the measurement of chemiluminescence. The reagent solution, situated in the lower part of the cuvette, contained a mixture of $0.7\,mM$ luminol and $3.8\,\mu M$ hemin, which emits luminescence on interaction with free radicals. The pH of the reagent solution was adjusted to $10-11$ by adding $11.8\,mM$ Na_2CO_3 prior to deaeration by bubbling with N_2 gas [27]. The solution which contained the free radical sources, H_2O_2, epinephrine, and the glucose-glucose oxidase system, was held in the upper part of the cuvette before starting the experiments. The antioxidants were admixed either with the reagent solution, in the lower part of the cuvette (separate part), or with the free radical source solution, in the upper part of the cuvette (synchronous part). A comparison of the results of experiments carried out both ways allows conclusions to be drawn on an eventual interaction of the antioxidant with the enzyme by binding to an active site.

When using H_2O_2 as a free radical source, the CH 402 and MTDQ-DA were incubated with luminol + hemin reagent solution, in order to avoid direct contact between H_2O_2 and the antioxidants before luminescence detection started. The concentration of H_2O_2 in the composed volume of the reagent and radical source solutions was $4.10^{-7}\,M$. Various concentrations of antioxidants were prepared in 100-μl volumes and were added to 1000 μl of the reagent solution.

When studying the interaction of silibinin with H_2O_2 the antioxidant was placed in the lower part, while H_2O_2 was placed in the upper part, of the cuvette. The H_2O_2 concentration was $2.2 \times 10^{-6}\,M$ in the total volume. The nonenzymatic system consisted of 1 ml luminol reagent and various concentrations of silibinin, the total volume being 1.15 ml.

The glucose-glucose oxidase system contained 92 μl $0.5\,M$ glucose in $0.01\,M$ acetate buffer, pH 5.0, and 8 μl glucose oxidase enzyme from $20\,mU/ml$ suspension [4]. Experiments were carried out with antioxidants held either in the reagent solution or the radical source solution prior to the measurement of luminescence. Various antioxidant concentrations were prepared in 100-μl volumes, the luminol + hemin reagent mixture comprising 1000 µl volume. The incubation time was 60 s in either the synchronous or the separate phase. The chemiluminescence intensity was expressed in millivolt seconds. The reaction times were 60 s.

The O_2^- scavenging effect of silibinin was examined with the chemiluminescence measurement method using autooxidation of epinephrine as the radical source. The concentration of epinephrine was $5.2 \times 10^{-5} M$ in a total volume of 1.15 ml. The reagent solution had a volume of 1 ml, contained antioxidants at various concentrations, and was held in the lower part of the cuvette, admixed to the luminol + hemin reaction mixture. Reaction times and the measurements taken can be seen in the figures.

Free radicals were detected in the fresh liver homogenates after 10 min incubation time at 37°C in separate phase. A solution of 200 µl luminol reagent was added to 1000 µl 25 w/v % tissue homogenate. The chemiluminescence intensity was expressed and recorded in millivolts.

The effects of CH 402, MTDQ-DA, and silibinin in in vivo treatments on the O_2^- scavenging function of fatty liver in rats were measured in a system containing epinephrine as the free radical source. The O_2^- radical source and tissue homogenates were in the synchronous part in the cuvette and luminol reagent was separated on measurement. The incubation time was 30 s. The measurements taken can be seen in the figures.

The effect of silibinin treatment on the H_2O_2-scavenging function of fatty liver was detected in experiments where the protein concentration was 1 mg/ml in 10 µl in the upper part of the cuvette, in synchronous part with the H_2O_2. The H_2O_2 concentration was $2.2 \times 10^{-6} M$ in the total volume and the volume of luminol reagent solution was 1 ml in the separate part in the cuvette. The incubation time was 15 s, with measurement times as in the figures. The temperature was 25°C during the experiments.

Statistical Analysis

Statistical analysis was performed by a two-tailed t test with a probability level of $p < 0.05$. Mean values represent the data obtained from five experiments.

Chemicals

Glucose-6-phosphate dehydrogenase and serum albumin were obtained from Calbiochem AG (Lucerne), acetyl acetone, cholic acid, luminol, hemin, NADPH, and NADH from Sigma (St Louis), glucose oxidase from Boehringer (Mannheim), total cholesterol, total lipids, and triglycerides and precipitating reagent for LDL and VLDL from Merck (Darmstadt), HDL and the total cholesterol kit from Baker (Egham), triglycerides from *SKI* (Vienna), and all other reagents from Reanal (Budapest).

Results

Histopathological investigation of the liver of the animals fed a fat-rich diet showed the following characteristic pathomorphological changes: fatty degeneration, balloon cell degeneration, centrilobular fatty degeneration or necrosis, and vacuolization. In comparison to normal liver structure (Fig. 1), at the periphery and sometimes the whole lobuli of the liver of the animals fed a fat-rich diet showed diffuse degenerative hepatocellular changes (Figs. 2, 3). As a result, the structure was somewhat foamy, the cell borders were indistinct, and some cells were rounded and vacuolated. However, no inflammatory reaction was observed.

In the electron microscopic investigations the liver cell cytoplasm presented an extremely variable appearance which reflected to some extent the functional state of the cell. The decrease in the glycogen content of the hepatic cells was considered to represent the first sign of liver injury. The glycogen was replaced by lipids in proportion to the decrease in glycogen content. In the cytoplasm large vacuoles could be seen (Figs. 4, 5). No glycogen was found in the zonal parts and the areas of fatty degeneration. An accumulation of various-sized lipid droplets up to 5 μm in diameter was observed in the cytoplasm of hepatocytes. At the low magnifications, these droplets appeared generally as spherical, electronlucent vacuoles, although some showed irregular shapes, and contained various inclusions in their

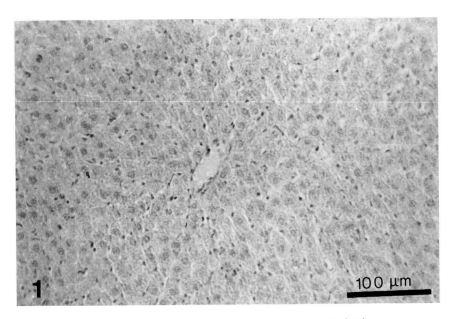

Fig. 1. Light microscopic picture of the liver tissue from control animals

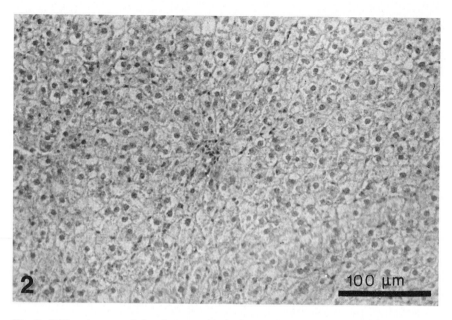

Fig. 2. Diffuse hepatocellular degeneration in liver from animals fed a fat-rich diet

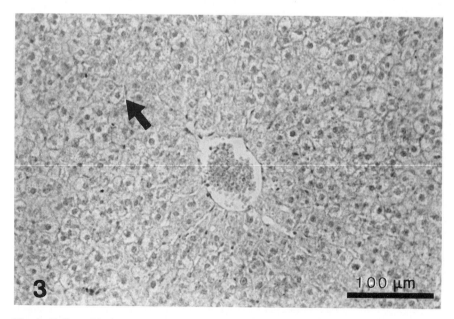

Fig. 3. Balloon-like hepatocytes (*arrow*) after an atherogenic diet

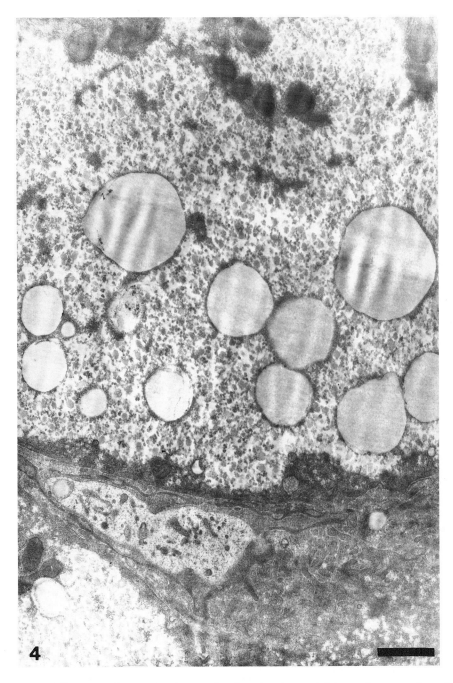

Fig. 4. Electron microscopic picture of a hepatocyte, containing a large number of electronlucent lipid droplets after an atherogenic diet

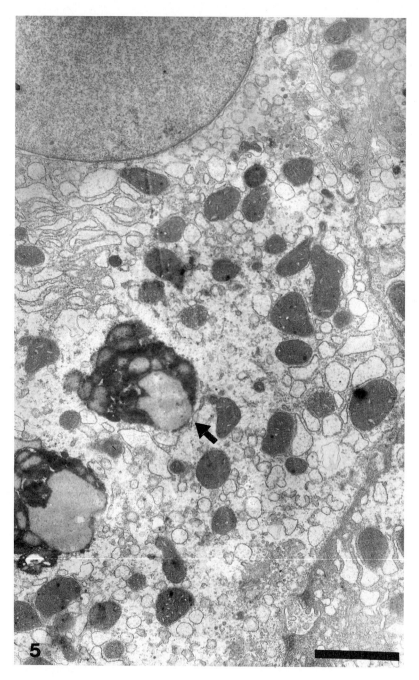

Fig. 5. Hepatocytic lysosomes (*arrow*) in the cytoplasm of liver from animals kept on a fat-rich diet

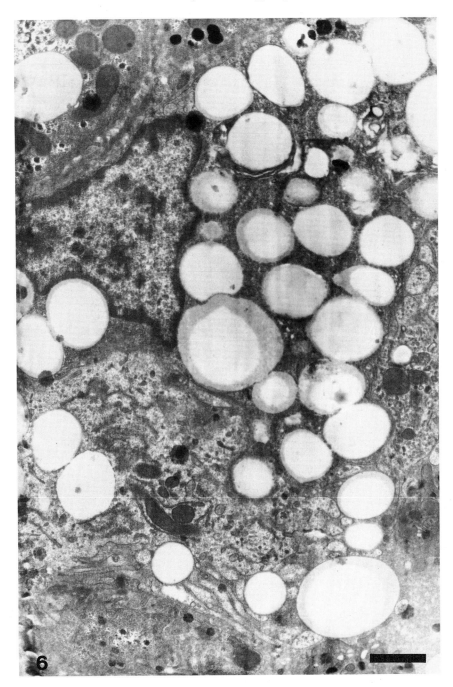

Fig. 6. Lipid-containing cell from the connective tissue of liver after an atherogenic diet

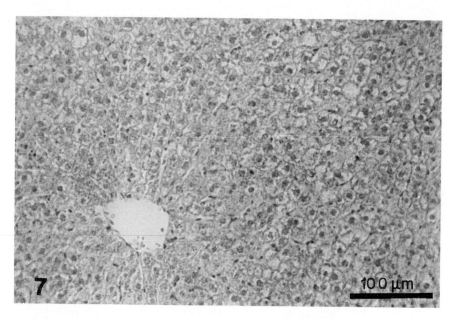

Fig. 7. After CH 402 treatment the centrilobular cells have a normal appearance, and the perilobular cells show alterations

periphery. The majority were seen to be enveloped by a unit membrane, approximately 6–8 nm thick, similar to the hepatocytic lysosomes, although there were a few lipid droplets lacking such a limiting membrane. Portions of the thickened layer usually contained heterogeneous substances, for instance, a myelin-like structure, electron-dense material, or electronlucent small droplets. These may be lipolysosomes (Fig. 5). A great number of lipid-containing cells were observed in the connective tissue (Fig. 6). The cell borders were separated, and the nuclei and nucleoli were normal.

After the CH 402 treatment only very small changes were observed. Some of the centrilobular cells showed a normal appearance (Fig. 7). However, after the MTDQ-DA and silibinin treatment the whole hepatic lobuli showed the structure of the liver to be similar to that in the controls (Figs. 8, 9). In the normal animals and following MTDQ-DA and silibinin treatment large numbers of characteristic aggregates consisting of glycogen and ribosome particles were seen in the hepatocytes using electron microscopy (Fig. 10). Some of those localized in the cytoplasm were partially surrounded by membranes of the endoplasmic reticulum.

A major factor in oxygen toxicity in aqueous solution is O_2^-, which can combine together with H_2O_2 in the presence of metal ions to generate the high reactive hydroxyl radical. The attack by oxygen-free radicals against the biological membranes initiates the process of lipid peroxidation. In this process the fatty acid side chains of the membrane lipids, especially those

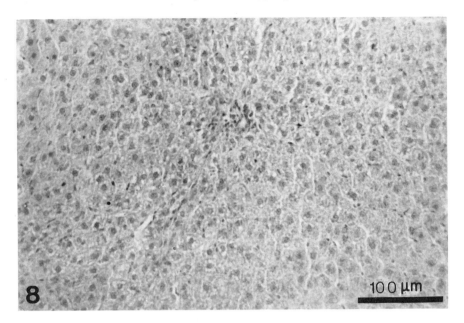

Fig. 8. In animals fed an atherogenic diet with MTDQ-DA treatment the structure of the liver is similar to that in the control

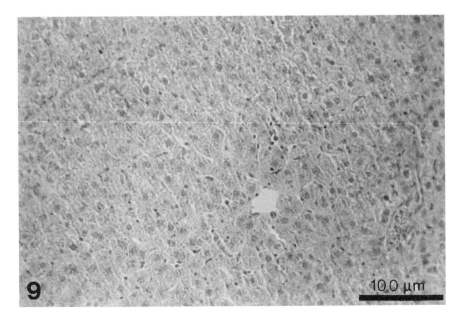

Fig. 9. After silibinin treatment, the liver structure shows a normal appearance in animals fed a fat-rich diet

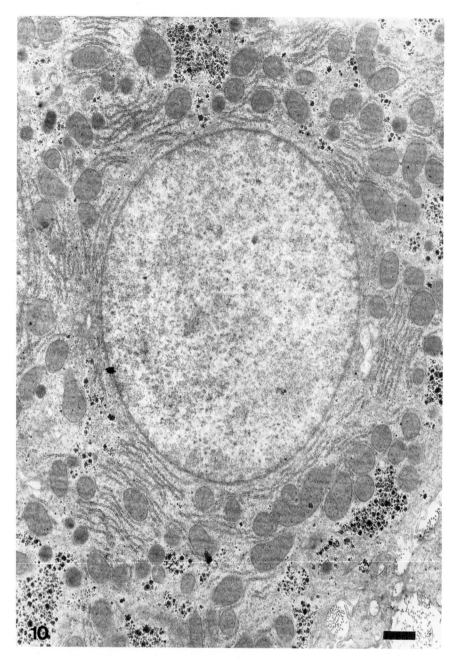

Fig. 10. Electron microscopic picture of a hepatocyte. A large quantity of glycogen can be observed in the cytoplasm after silibinin treatment and a fat-rich diet

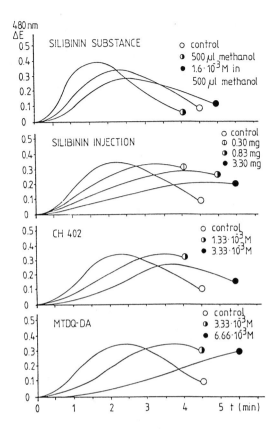

Fig. 11. Effect of antioxidants on the autooxidation of epinephrine

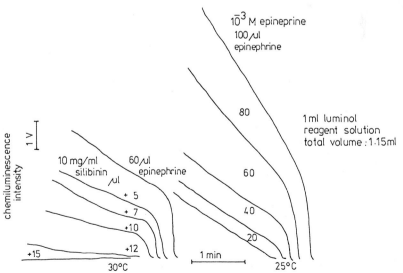

Fig. 12. O_2^--scavenging effect of silibinin injection in vitro

containing two or more carbon-carbon double bonds, are oxidized to hydroperoxides. In the presence of metal catalyst, hydroperoxides decompose to form a complex mixture of hydrocarbons and cytotoxic aldehydes, for example malondialdehyde. In general terms, natural and synthetic antioxidants are compounds that inhibit lipid peroxidation by interfering with the chain reaction of peroxidation and/or by scavenging reactive oxygen radical and/or by repairing damaged sites of biopolymers caused by free radicals.

We investigated the antioxidant properties of CH 402, MTDQ-DA, and silibinin in further various systems in which oxygen free radicals were generated in enzymatic or nonenzymatic pathways.

The autooxidation of epinephrine is a nonenzymatic source of superoxide radical ions (O_2^-), an intermediate species in the multistep process. The absorbance at 480 nm due to the product of the reaction, adrenochrome, is a function of reaction time in the absence and presence of antioxidants. The delayed formation of adrenochrome and the reduced absorbance found when antioxidants were present can be attributed to the scavenging of O_2^-.

The curves show the dose-dependent inhibitory effect of MTDQ-DA, CH 402, and silibinin. Silibinin was dissolved in methanol; therefore methanolic control was also measured. The data are presented in Fig. 11.

The production of light from chemical emanates from the excited state. Chemiluminescence is a property of excited states of luminol formed in a medium containing free radicals O_2^-, $\cdot OH$, or their precursor H_2O_2. In this experiment to measure chemiluminescence, the scavenging effect of silibinin on O_2^- was studied (Fig. 12). The results confirm that the silibinin injection solution reacts with O_2^- anion during autooxidation of epinephrine.

Hydrogen peroxide is able to induce free radical reactions with organic compounds, presumably via $\cdot OH$ radicals. Figure 13 shows the scavenging effect of silibinin injection solution at increasing concentrations on the H_2O_2 molecule. The concentration of H_2O_2 was measured in control experiments. The antioxidant was also in the lower part of the cuvette in separate phase. The data in Table 1 show a progressive decrease of luminescence when

Table 1. Scavenger effect on H_2O_2 (4×10^{-7} M) by CH 402 and MTDQ-DA: variation of chemiluminescence intensity (mV s) with concentration of antioxidant

Concentration (M)	Antioxidant	
	CH 402	MTDQ-DA
0	57.511	57.511
8.7×10^{-7}	53.468	55.708
8.7×10^{-6}	30.480	37.609
8.7×10^{-5}	28.587	7.279
8.7×10^{-4}	0.001	0.530

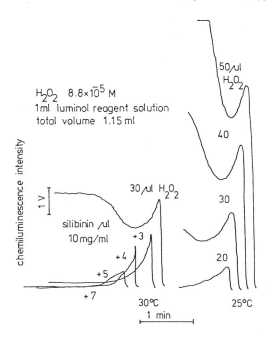

Fig. 13. H_2O_2-scavenging effect of silibinin injection in vitro

Table 2. Scavenger effect on H_2O_2 by CH 402 and MTDQ-DA in the glucose-glucose oxidase system: variation of chemiluminescence intensity (mV s) with concentration of antioxidant

In separate part for 60 s			In synchronous part for 60 s		
Concentration $(M)^a$	Antioxidant		Concentration $(M)^b$	Antioxidant	
	CH 402	MTDQ-DA		CH 402	MTDQ-DA
0	47.693	47.693	0	47.693	47.693
8.3×10^{-7}	28.802	31.197	5×10^{-6}	33.165	30.768
8.3×10^{-6}	12.755	25.974	5×10^{-5}	27.222	28.420
8.3×10^{-5}	8.545	3.925	5×10^{-4}	22.403	22.420
8.3×10^{-4}	0.672	0.513	5×10^{-3}	0.615	0.273

[a] Concentrations of antioxidant in the total volume.
[b] Concentrations of antioxidant in the upper part of the cuvette.

increasing amounts of CH 402 or MTDQ-DA were added, the effect being attributed to the scavenger activity of the compounds. In these experiments the antioxidants had been incubated with the luminol + hemin reagent solution, separated from H_2O_2, before detection of chemiluminescence intensity, in order to prevent them from directly interacting. Table 2 shows the scavenging effect of CH 402 and MTDQ-DA on H_2O_2 generation in the glucose-glucose oxidase system in synchronous and separate parts. The

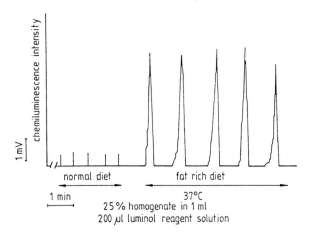

Fig. 14. Free radical level in fatty liver

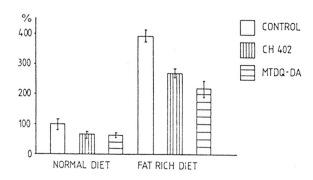

Fig. 15. Conjugated diene levels of rat liver homogenates

incubation time of the enzyme reaction as well as the time of luminescence detection was 60 s in these experiments.

We also examined the biochemical changes in liver of hyperlipidemic rats to assess to what extent the free radical reactions took part in the fatty degeneration. It can be seen that the chemiluminescence of control liver homogenates was not detectable or it only had a very weak intensity. Fatty livers of rats fed on an atherogenic diet resulted in a considerable chemiluminescence intensity as shown in Fig. 14.

Change of free radical level in experimental steatosis could be justified by ultraviolet spectrophotometric detection of conjugated dienes (Fig. 15), which appeared in peroxidized unsaturated fatty acids following free radical attack. The antioxidant treatments inhibited the reaction, leading to the formation of conjugated double bonds. These results suggest some pathological changes of the animal tissue induced by the atherogenic diet, which may be a lipid peroxidative fatty degradation.

It has been known that the redox transformation of iron promotes lipid peroxidation. Reducing enzymes such as NADPH cytochrome 450 reductase

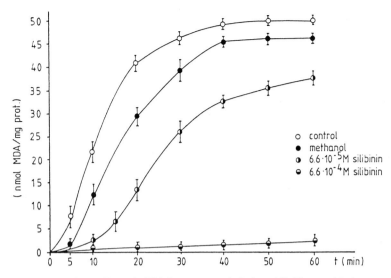

Fig. 16. In vitro effect of silibinin on enzymic-induced lipid peroxidation

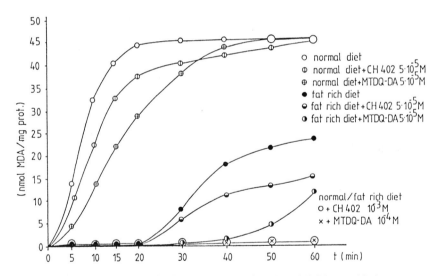

Fig. 17. In vitro effect of antioxidants on enzymic-induced lipid peroxidation

can reduce iron complexes. During NADPH oxidation liver microsomes produce significant amounts of $\cdot OH$ radicals, which are known to be more active in inducing lipid peroxidation than superoxide radical ions. The NADPH + Fe^{3+}-induced lipid peroxidation was studied in vitro using liver microsomal fraction. The CH 402, MTDQ-DA, and silibinin re-

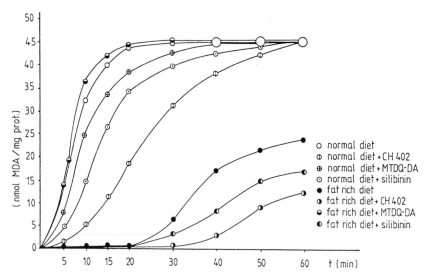

Fig. 18. Effect of in vivo antioxidant treatments on enzymic-induced lipid peroxidation

duced the induced lipid peroxidation in both cases depending on time and concentration.

The CH 402, MTDQ-DA, and silibinin did not affect the activity of enzymes, NADPH cytochrome P-450 reductase, and polysubstrate monooxygenase system in vitro and in vivo to any significant extent, as the following data show. The extent of NADPH-induced and Fe^{3+}-stimulated lipid peroxidation was detected by monitoring MDA production. Figure 16 shows, that adding various concentrations of silibinin to the system in methanol medium caused a significant lowering of the MDA level compared to the control. A quantity of $6.6 \times 10^{-4} M$ silibinin added in the same way, resulted in an almost complete inhibition of malondialdehyde production. Figure 17 shows that CH 402 and MTDQ-DA completely inhibited the enzymatically induced lipid peroxidation.

The NADPH + Fe^{3+}-induced lipid peroxidation was studied in vitro on liver microsome preparations in normo- and hyperlipidemic rats. Considerable differences were detected between the two groups in the production of MDA during lipid peroxidation. The lipid peroxidation was also inhibited by CH 402, MTDQ-DA, and silibinin in samples from animals kept on a fat-rich diet.

The in vivo CH 402 and silibinin treatments inhibited the in vitro enzymatically induced lipid peroxidation both in control and hyperlipidemic rat liver microsomes.

The in vivo MTDQ-DA treatment, in contrast to two other antioxidants, stimulated the in vitro induced process in the sample, which was prepared from hyperlipidemic rat livers (Fig. 18). MTDQ-DA increased the

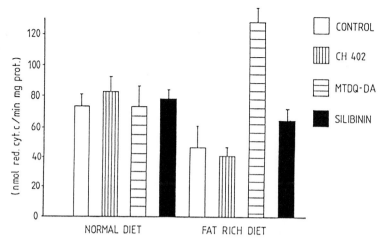

Fig. 19. Effect of in vivo antioxidant treatments on NADPH cytochrome c reductase activity in normo- and hyperlipidemic rats

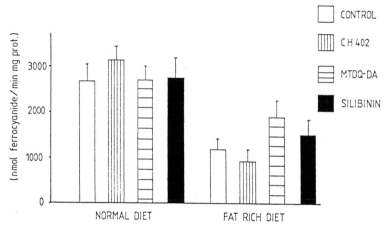

Fig. 20. Effect of in vivo antioxidant treatments on NADH ferricyanide reductase activity in normo- and hyperlipidemic rats

activity of NADPH cytochrome c reductase of liver microsomal fractions obtained from hyperlipidemic rats to a greater extent than that of rats fed on a normal diet (Fig. 19).

CH 402 and silibinin had no significant effect on the enzyme activity both in control and in treated samples, but silibinin increased the rate of activity nonsignificantly in this system.

The effect of CH 402, MTDQ-DA, and silibinin on NADH ferricyanide reductase enzyme activity in the hyperlipidemic rats is shown in Fig. 20.

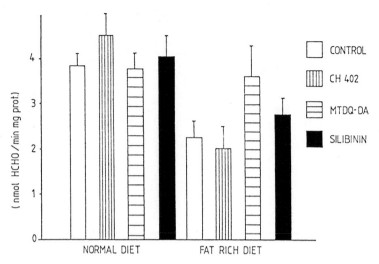

Fig. 21. Effect of in vivo antioxidant treatments on the activity of *N*-demethylase in normo- and hyperlipidemic rats

Table 3. Cytochrome P-450 concentration following anti-oxidant treatment

Diet	Concentration (nmol/mg protein)
Normal diet	0.533 ± 0.052
Normal diet + CH 402	0.512 ± 0.063
Normal diet + MTDQ-DA	0.528 ± 0.048
Normal diet + silibinin	0.561 ± 0.071
Fat-rich diet	0.353 ± 0.079
Fat-rich diet + CH 402	0.310 ± 0.085
Fat-rich diet + MTDQ-DA	0.466 ± 0.082
Fat-rich diet + silibinin	0.510 ± 0.078

Each figure represents 25 different rats.

MTDQ-DA strongly increased the enzyme activity in microsome isolated from hyperlipidemic rat livers. CH 402 and silibinin also had similar effects in this system.

MTDQ-DA had a significant effect on the activity of *N*-demethylase in hyperlipidemic rat microsomal fraction, and CH 402 and silibinin treatment also had no significant effects in the experiment, but silibinin increased, and CH 402 lowered, the enzyme activity (Fig. 21). The antioxidants did not increase cytochrome P-450 content when the rats were kept on a normal diet, but MTDQ-DA increased the cytochrome P-450 concentration considerably when the rats were fed on a fat-rich diet (Table 3).

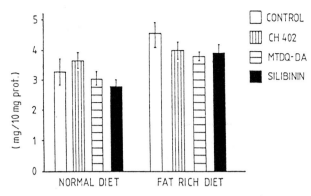

Fig. 22. Changes in lipid parameters of liver microsomal membranes of hyperlipidemic rats during antioxidant treatment (total lipids)

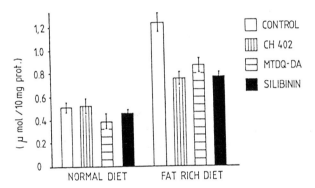

Fig. 23. Changes in lipid parameters of liver microsomal membranes of hyperlipidemic rats during antioxidant treatment (total cholesterol)

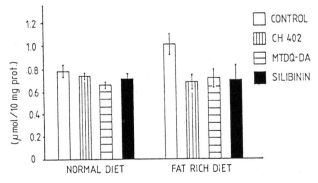

Fig. 24. Changes in lipid parameters of liver microsomal membranes of hyperlipidemic rats during antioxidant treatment (triglycerides)

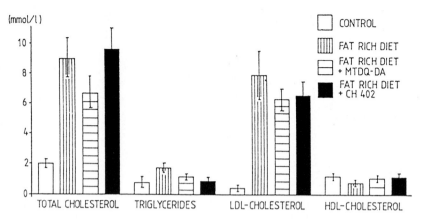

Fig. 25. Effect of dihydroquinoline-type antioxidants on lipid parameters of sera in hyperlipidemic rats

The lipid parameters were also changed in the liver microsomes during the antioxidant treatments, but the mechanisms are unknown as yet. Presumably, a membrane protective effect was expressed in the data. Figures 22–24 show that the antioxidant treatments did not change the lipid parameters – total lipids, total cholesterol, and triglycerides – in normolipidemic rat liver microsomes, but the values were reduced by CH 402, MTDQ-DA, or silibinin administration in steatotic liver microsomes in short-term experiments. Our experiments showed that the lipid peroxidations of sera were increased during fattening, and both MTDQ-DA and CH 402 had some lowering effect on total and LDL cholesterol and on serum triglycerides in rats kept on an atherogenic diet; HDL cholesterol was not affected (Fig. 25).

Living organisms are protected against reactive oxygen intermediates and lipid peroxides by certain natural antioxidants (vitamins, tiol, -amine groups, bioflavonoids), various enzymes (superoxide dismutase, catalase, peroxidase, glutathione peroxidase), and the P-450 system to decompose the end products of lipid peroxidation or xenobiotics.

The natural scavenger capacity of tissue is changed according to the maximum curve. If the defense mechanisms of living cells are badly damaged, the total free radical scavenger activity is decreased.

We would like to demonstrate in Fig. 26 and Table 4 that the protecting activity against O_2^- and $H_2O_2/\cdot OH$ becomes stronger under antioxidant treatments in rat liver. Although the specific enzyme activities could not be measured by these chemiluminescence measurement techniques, the data demonstrated that the antioxidants we used had a membrane-protecting effect, and increased the natural total scavenger capacity (Fig. 26, Table 4).

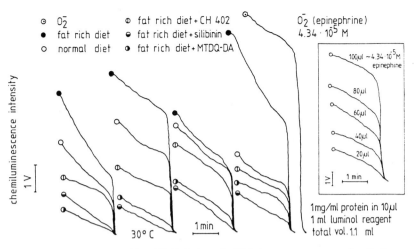

Fig. 26. Effect of CH 402 MTDQ-DA and silibinin treatments on the O_2^--scavenging function of fatty liver in rats

Table 4. Effect of silibinin on the H_2O_2-scavenging function of fatty liver in rats

Samples ($n = 10$)	Chemiluminescence intensity (mV s) ($\bar{x}$)
H_2O_2	1 313 000
Normal diet	124 412
Normal diet + silibinin	65 430
Fat-rich diet	283 675
Fat-rich diet + silibinin	208 062

Incubation time, 15 s; reaction time, 60 s; temperature, 25°C; total volume, 1.15 ml; concentration of H_2O_2, $8.8 \times 10^{-5} M$ in 50 µl; luminol reagent solution, 1 ml.

Discussion

The present study revealed some biochemical and morphological changes caused by free radical reactions in experimental hyperlipidemia in the pathomechanism of fatty liver. We investigated the role of free radical reactions, lipid peroxidation, and the protecting effect of some antioxidants in steatosis hepatitis. By electron microscopy a large number of lipolysosomes were observed in the hepatocytes. After administration of cholesterol or egg yolk, a proliferation of hepatocytic lipolisosomes has been reported [25, 26, 30, 31]. Steatotic liver, reaching peak lipid accumlation, shows declining activity and inductibility of enzymes. On the basis of our present experiments we confirmed that the free radical level rose and natural scavenger capacity decreased significantly in the fatty liver. The antioxidants supported the natural scavenger capacity, consequently decreasing the risk of atherosclerosis in hyperlipidemia.

According to Freeman and Crapo, the cellular sources of free radicals are mainly plasma membranes, endoplasmic reticulum, nuclear membrane, peroxisomes, and small molecules. And the reactive oxygen intermediates may be formed in all parts of the cell [19].

The integrity of cell membrane functions and enzyme activities by dihydroquinoline and flavonolignane-type derivatives have been studied in various in vitro and in vivo free radical generating systems by spectrophotometry and luminescence measurement techniques [4, 17].

It was justified that these chemicals could combine with O_2^- anions and H_2O_2 and in this way inhibit the oxygen-stimulated lipid peroxidation [22, 37]. The reactivity of MTDQ-DA, CH 402, or silibinin in capturing superoxide ions and $\cdot OH$ radicals was confirmed in this study using chemical sources of radicals ($H_2O_2/\cdot OH$, epinephrine/O_2^-).

The results obtained with the enzymatic glucose-glucose oxidase source can be explained with the same kind of scavenging activity of the compounds.

The substances MTDQ-DA, CH 402, or silibinin might, however, blockade some sites of the oxidase, among them those responsible for the production of H_2O_2. The luminescence measurement data summarized in the tables argue against this since they indicate that the enzymes preserved their activity in producing free radicals. Binding of the antioxidants can greatly influence the mechanism of the free radical induced reactions in biological systems, which should be discovered in further experiments. Work is in progress in this field. Furthermore, we detected the membrane protective effect of the chemicals we used in rat liver microsomal membranes in in vitro experiments. Inhibition on both enzymatic and nonenzymatic induced lipid peroxidation of CH 402, MTDQ-DA, and silibinin have been verified.

Decreased hepatic reduced glutathione (GSH) and increased conjugated diene levels were found in biopsy specimens of the liver of patients with alcoholic liver disease and acute carbon tetrachloride poisoning [43, 45]. Both acute and chronic alcoholic administration to rats and baboons elevated the conjugated diene level and significantly depressed the GSH level in the liver [44]. We detected a reduction of diene conjugation under antioxidant treatments in experimental hyperlipidemia in rat liver.

It was shown that none of the antioxidants affected the activity of enzymes NADPH cytochrome P-450 reductase and the polysubstrate monooxygenase system in vitro or in vivo to a significant extent in normolipemic controls.

These derivatives lowered the induced lipid peroxidation in vitro, but the two dihydroquinoline-type derivatives showed different effects in hyperlipidemic rats in vivo.

If the lipid peroxidation was induced by NADPH and Fe^{3+} in vitro, the in vivo MTDQ-DA treatment increased the enzyme activity and the lowered value was normalized.

The substance MTDQ-DA increased the enzyme activity of NADPH cytochrome c reductase, NADH ferricyanide reductase, and N-demethylase and cytochrome P-450 in vivo in hyperlipidemic rats. CH 402 did not significantly increase enzyme activity either in normo- or in hyperlipidemic rats. Silibinin treatment had a moderate increasing effect on microsomal enzyme activity in hyperlipidemia.

On the bases of these results, our hypothesis is that MTDQ-DA besides its chemical properties has a multicentric effect, among others membrane protecting, antioxidant, and scavenging – so that its protective activity against membrane destruction in microsomes in hyperlipidemia is stronger than that of the other two antioxidants (CH 402 and silibinin) – consequently the condition of the liver during fatty degeneration is better than during treatment with CH 402 or silibinin for the same period.

The antioxidants CH 402 and MTDQ-DA had some lowering effect on serum lipids, on total and LDL cholesterol, and on the serum triglycerides in rats kept on an atherogenic diet. HDL cholesterol was not affected in dihydroquinoline-type chemical treatments.

The antioxidant treatment lowered the total lipid, total cholesterol, and triglyceride levels in the microsomal membranes at different rates in normo- and in hyperlipidemic rats. It was established that antioxidants had lipid-lowering effects in "short-term" experiments; thus it can be presumed that these chemicals influence the pathological lipid metabolism in different ways at different rates.

Strong synthetic antioxidants cannot be used in human therapy because of their toxicity in effective doses (15). Our results nevertheless support the conclusion that the administration of MTDQ-DA, CH 402, or silibinin has a favorable effect on several diseases in which oxidative stress, due in part to oxygen free radicals, is a primary factor in the pathogenesis.

Acknowledgements. The authors wish to express their thanks to Professor Ágnes J.N. Zsinka for her cooperation; to Miss Mária Miskolczi Török for detection of lipid parameters in the sera; to Mrs. Éva Nagy, Zsuzsanna Nagy, and Gizella Bacsó of the Arteriosclerosis Research Group and Mrs. Györgyi Tóth of the First Department of Anatomy for their excellent technical assistance.

References

1. AOAC Official methods of analysis (1984)
2. Bindoli A, Cavallini L, Siliprandi N (1977) Inhibitory action of silymarin on lipid peroxide formation in rat liver mitochondria and microsomes. Biochem Pharmacol 26:2405–2409
3. Blázovics A, Somogyi A, Lengyel G, Láng I, Fehér J (1988) Inhibition of lipid peroxidation by dihydroquinoline type antioxidant (CH 402). Free Rad Res Comms 6:409–413

4. Blázovics A, György I, JN Zsinka A, Biacs P, Földiák G, Fehér J (1989a) In vitro scavenger effect of dihydroquinoline type derivatives in different free radical generating systems. Free Rad Res Comms 4:217–226

5. Blázdvics A, Somogyi A, Ambrus D, Mathiász D, Vereckei A, Fehér J (1989b) The effect of CH 402 dihydroquinoline type antioxidant on the activity of Na^+K^+-ATP-ase and Mg^{++}-ATP-ase of rat brain subcellular fractions in the presence and absence of ascorbic acid. Acta Physiol Hung 73:3–7

6. Cavallini L, Bindoli A, Siliprandi N (1978) Comparative evaluation of antiper-oxidative action of silymarin and other flavonoids. Pharmacol Res Commun 10: 133–138

7. Csomós G, Thaler H (1983) Clinical hepatology. History, present state, outlook. Foreword: Popper H. Springer, Berlin Heidelberg New York

8. Csomós G, Hruby K, Thaler M (1985) Silibinin in the treatment of deathcap fungus poisoning. Q Bull Hung Gastroenterol Soc 3:39–46

9. Del Maestro RF (1980) An approach to free radicals in medicine and biology. Acta Physiol Scand Suppl 492:153–168

10. Demopoulus HB, Flamm ES, Pietronigro D, Seligman ML (1980) The free radical pathology and the microcirculation in the major central nervous system disorders. Acta Physiol Scand Suppl 492:91–119

11. Dormandy TL (1978) Free radical oxidation and antioxidants. Lancet: 647–650

12. Fehér J, Bär-Pollák Zs, Sréter L, Fehér E, Toncsev H (1982) Biochemical markers in carbon tetrachloride and galactosamine-induced acute liver injuries: the effect of dihydroquinoline-type antioxidants. Br J Exp Pathol 63:394–400

13. Fehér J, Sulyok S, Pollák Zs, Toncsev H, Cornides Á, Blázovics A, Szondy É, Gerö S (1984) The effect of a recently developed dihydroquinoline-type radical scavenger in cholesterol induced hyperlipidemia. In: Lenzi S, Descovich GC (eds) Arterio-sclerosis and cardiovascular diseases. Compositori, Bologna, pp 87–91

14. Fehér J, Kiss A, Blázovics A, Szondy É, Toncsev H, Mathiász D, Gerő S (1985) Hypolipidaemic effect and inhibition of lipid peroxidation with glunicate in rats treated with atherogenic diet. Drug Exp Clin Res 11:413–419

15. Fehér J, Csomós G, Vereckei A (1987a) Free radical reactions in medicine. Springer, Berlin Heidelberg New York

16. Fehér J, Láng I, Nékám K, Csomós G, Müzes Gy, Deák G (1987b) Effect of silibinin on the activity and expression of superoxide dismutase (SOD) in lymphocytes from patients with chronic alcoholic liver diseases. Free Rad Res Comms 6:373–376

17. Fehér J, Blázovics A, György I, Vereckei A, Somogyi A, Cornides A (1989) The effect proved by pulse radiolysis and chemiluminometric methods of free radical scavengers in liver lesions. EASL J Hepatol [Suppl 1]:9 S148

18. Fiebrich F, Koch H (1979) Silymarin an inhibitor of lipoxygenase. Experientia 35:1548–1560

19. Freeman BA, Crapo JD (1982) Biology of disease: free radicals and tissue injury. Lab Invest 47:412–426

20. Fridovich I (1987) The biology of oxygen radicals: general concepts. Upjohn Symposium/Oxygen radicals pp 1–5

21. Friedewald WT, Lewy RI, Fredrickson DS (1972) Estimation of the concentration of LDL-cholesterol in plasma without use of the preparative ultracentrifuge. Clin Chem 18:499–506

22. Gutteridge JM (1987) Lipid peroxidation: some problems and concepts. Upjohn Symposium/Oxygen radicals: pp 1–19

23. Halliwell B, Gutteridge JM (1984) Lipid peroxidation, oxygen radicals, cell damage and antioxidant therapy. Lancet i:1936–1937

24. Harman D (1981) The aging process. Proc Natl Acad Sci USA 78:7124–7128

25. Hayashi H, Winship DH, Sternlieb, I (1977) Lipolysosomes in human liver: dis-tribution in livers with fatty infiltration. Gastroenterology 73:651–654

26. Hayashi H, Sameshima Y, Lee M, Hotta Y, Kosaka T (1983) Lipolysosomes in human hepatocytes: their increase in number associated with serum level of cholesterol in chronic liver diseases. Hepatology 3:221–225

27. Heide L, Bögl, W (1986) The identification of irradiated dried food-stuffs by luminescence measurements. Food Lab Newslett 5:21–23
28. Hornsby PJ, Crivello JF (1983) The role of lipid peroxidation and biological anti-oxidants in the function of the adrenal cortex. Part I. A background review. Mol Cell Endocrinol 30:1–20
29. Jansson J, Schenkman JB (1977) Studies on three microsomal electron-transfer enzyme systems (specificity of electron flow pathways). Arch Biochem Biophys 178:89–107
30. Kanai M (1989) Ultrastructural and biochemical studies of lipolysis by lipolysosomes in chick hepatocytes. Cell Tissue Res 255:559–565
31. Lee M, Hayashi H, Kato S, Sameshima Y, Hotta Y (1982) Egg yolk-induced lipolysosome proliferation and fat infiltration of rat liver. Lab Invest 47:194–197
32. Lowry AH, Rosenbrough NJ, Farr AL, Randall RJ (1951) Protein measurement with the Folin-phenol-reagents. J Biol Chem 193:265–275
33. Marklund SL, Westman NG, Lundgren E, Ross G (1982) Copper and zinc-containing superoxide dismutase, manganese-containing superoxide dismutase, catalase, and glutathione peroxidase in normal and neoplastic human cell lines and normal human tissues. Cancer Res 42:1955–1961
34. Meins R, Heinrich V, Robenev H, Themen H (1982) Effect of silibinin on hepatic cell membranes after damage by polycyclic aromatic hydrocarbons. Agents Action 12:254–257
35. Misra HP, Fridrovich I (1972) The role of superoxide anion in the autooxidation of epinephrine and a simple assay for superoxide dismutase. J Biol Chem 247:3170–3175
36. Nash T (1953) Colorimetric estimation of formaldehyde by means of the Hantzsch reaction. Biochem J 55:416–421
37. Nizzamuddin A (1987) NADPH dependent and O_2^- dependent lipid peroxidation. Biochem Educ 15:58–63
38. Omura T, Sato R (1964a) The carbon monoxide-binding pigment of liver micro-somes. J Biol Chem 239:2370–2378
39. Omura T, Sato R (1964b) The carbon monoxide-binding pigment of liver micro-somes. J Biol Chem 239:2379–2385
40. Ottolenghi A (1959) Interaction of ascorbic acid on mitochondrial lipids. Arch Biochem Biophys 79:355–363
41. Pryor WA (1973) Free radical reactions and their importance in biochemical systems. Fed Proc 32:1862–1869
42. Pryor WA (1982) Free radical biology, xenobiotics, cancer and aging. Ann N Y Acad Sci 393:1–22
43. Ruprah M, Mant TGK, Flanagan RJ (1985) Acute carbon tetrachloride poisoning in 19 patients: implications for diagnosis and treatment. Lancet i:1027–1029
44. Shaw S, Jayatilleke E, Ross WA, Gordon EF, Lieber CS (1981) Ethanol-induced lipid peroxidation potentiation by long-term alcohol feeding and attenuation by methionine. J Lab Clin Med 98:417–424
45. Shaw S, Rubin KP, Lieber CS (1983) Depressed hepatic glutathione and increased diene conjugates in alcoholic liver disease. Evidence of lipid peroxidation. Dig Dis Sci 28:585–589
46. Toncsev H, Pollák Z, Kiss A, Fehér J (1982) Acute carbon tetrachloride induced lysosomal membrane damage and the membrane protecting effect of a new di-hydroquinoline-type antioxidant. Int J Tissue React 4:325–330
47. USA Patent, Nr. 4356306 (MTDQ-DA)
48. USA Patent, Nr. 4363910 (CH 402)
49. Valenzuela A, Guerra KI (1986) Differential effect of silybin on the Fe^{2+}-ADP and t-butyl-hydroperoxide induced microsomal lipid peroxidation. Experientia 42:139–141
50. Zsinka AJN, Blázovics A, Biacs P (1988) Chemiluminescence phenomena in animal tissues of different quality. Hung Sci Instrum 64:11–13

Free Radical Reactions in the Pathomechanism of Amiodarone Liver Toxicity

A. Vereckei, E. Fehér, A. Blázovics, J. György, H. Toncser, and J. Fehér

Introduction

Amiodarone (AMI) is one of our most potent antiarrhythmic drugs, with special chemical, pharmacokinetic, and electrophysiological properties. It is effective in arrhythmias originating from every part of the myocardium, applicable in grave, perilous rhythm disturbances (e.g., refractory ventricular tachycardia, atrial fibrillation with a high ventricular rate associated with anterograde Kent bundle conduction, and refractory supraventricular tachycardia). AMI treatment is the only therapeutic approach capable of preventing or at least significantly reducing the incidence of sudden death in hypertrophic obstructive cardiomyopathy in low doses. In survivors of sudden arrhythmic death – the group which is considered to be the most rigorous test of an antiarrhythmic agent – AMI treatment was effective in the significant reduction of mortality in contrast to the failure of conventional antiarrhythmic drugs. A further great advantage of AMI compared to other antiarrhythmic agents is that it has only insignificant proarrhythmic activity, and can be safely administered to patients with decreased left ventricular function [27, 28, 11].

The chronic administration of AMI is limited by the not infrequent and sometimes serious side effects. There are cardiac and extracardiac side effects (see Table 1). As Table 1 shows, AMI may damage practically every organ. In 0.5%–22% of patients such side effects were reported, which led to the cessation of AMI treatment, and this proportion shows a constant increase with the time of treatment. The pulmonary, cardiac, and liver involvement may result in the most serious consequences. Pulmonary toxicity, when clinically manifest, was lethal in 10%–20% of cases! That is why AMI in spite of its excellent antiarrhythmic activity can be administered only when other antiarrhythmic agents have failed in the treatment of the given rhythm disorder, when the latter is dangerous, and when the effectiveness of AMI treatment has been proved by electrophysiological and/or Holter examination. The most dangerous pulmonary, cardiac, and liver toxicity shows no relationship with maintenance, cumulative dose, or serum level of the drug [25–27, 42]. The pathogenesis of side effects has not been clarified, there are no appropriate screening tests for their prevention as yet, and their causal therapy in addition to preventive treatment and dosage

Table 1. Cardiac and extracardiac side effects

Cardiac complications
Bradycardia
Arrhythmogenic activity
Exacerbation or precipitation of heart failure
Conduction disturbances
Extracardiac side effects
Respiratory function alterations
Pneumonitis
Interstitial fibrosis of the lung
Bronchial asthma exacerbation
Elevated transaminase values
Hepatitis
Gastrointestinal intolerance
Photosensitivity
Slate gray discoloration of the skin
Corneal microdeposits
Visual disturbances
Thyroid function alterations (hyper- and hypothyreosis)
Peripheral neuropathy
Renal insufficiency

reduction has not been elucidated. Understanding the pathogenesis of side effects could be an essential step for this agent, which has almost ideal properties to become a first-line antiarrhythmic drug.

Indirect and direct mechanisms have been supposed to play a role in the pathomechanism of AMI toxicity. There are some data for the indirect, immunological mechanism. These are positive cutaneous and basophil degranulation tests with AMI, and in the bronchoalveolar lavage of patients with pulmonary toxicity some authors found lymphocyte accumulation and the inversion of normal proportion of helper/suppressor T cells, which was not confirmed by other authors (characteristic to hypersensitive pneumonitis), and positive lymphocyte transformation in the presence of AMI [27, 42]. In the organs of patients and experimental animals treated with AMI, lamellated lysosomal inclusions were found, which are probably composed of accumulated intralysosomal phospholipids. In the induction of generalized lysosomal phospholipoidosis the so-called amphiphilic property (a molecule which contains equal polar and nonpolar moieties) of AMI is involved; other drugs with similar a amphiphilic property, e.g., chloroquine, tilorone, perhexiline maleate, and tricyclic antidepressants, are also capable of inducing generalized phospholipoidosis. Generalized lysosomal phospholipoidosis can be explained on the basis of inhibition of phospholipase activity by amphiphilic compounds. They easily penetrate the cell membrane due to their amphiphilic property, accumulate in lysosomes, and form complexes with lipids stored in the lysosome. This change in the lipid substrate results in inhibition of lysosomal phospholipase activity. Besides, AMI also has an in vitro phospholipase inhibitory action.

Lysosomal phospholipoidosis is a specific marker of treatment with AMI or other amphiphilic drugs, some authors proposing that lysosomal phospholipoidosis may have an intimate role in the pathogenesis of AMI-induced generalized organ damage.

Other authors' findings have not supported this theory, because they found lysosomal phospholipoidosis in the organs of every patient treated with AMI independently of whether the given patient had any signs of toxic manifestation. On the basis of these data they concluded that lysosomal phospholipoidosis had no role in the pathogenesis of the side effects of amiodarone [14, 37]. We agree with the latter view in that the presence of lysosomal phospholipid inclusions cannot be considered a sign of drug toxicity, but we do not dismiss the pathogenetic role of intralysosomal phospholipid accumulation. We suppose that lysosomal phospolipid accumulation, reaching a certain level, becomes harmful to the cell, due to the direct cytotoxic and detergent activity of phospholipids, and in that way it may be involved in the pathogenesis of AMI toxicity.

Another potential mechanism of direct toxicity – which may also be related to lysosomal phospholipoidosis – is the involvement of pathological free radical reactions. In the voluminous AMI literature there are few communications dealing with the pathogenesis of side effects, and within these even fewer propose the role of free radical reactions. The communications already published are without exception based on in vitro studies, mainly dealing with the supposed free radical mechanism of AMI-induced photosensitivity. As the total number of the above-mentioned articles is five, we briefly summarize their results. Hassan and coworkers demonstrated that AMI induced the photohemolysis of red cells, which was partially oxygen dependent, and this effect was probably mediated by reactive oxygen intermediates [18]. Li and Chignell studied the photolysis of AMI and desethylamiodarone (DEAMI), and found that UV irradiation resulted in the formation of an aryl radical after deiodination, which induced lipid peroxidation in the presence of linoleic acid. In the presence of air its reaction with O_2 resulted in the secondary production of superoxide radicals ($O_2^{\cdot -}$). They explained by these processes the accumulation of lipofuscin in the skin of AMI-treated patients suffering from photosensitivity, concluding that photosensitivity was caused in that case by a phototoxic reaction [23]. Paillous and Verrier confirmed the above results in that they found the deiodination of AMI and DEAMI to be accompanied by aryl radical formation. They also identified the further reaction products. Before deiodination an intermediate excited triplet state of AMI was demonstrated and, in the presence of air, after the formation of the same excited triplet state, the formation of singlet oxygen ($^1O_2$) was observed from AMI [31]. On the other hand, Bennett and coworkers supposed that the carbonyl oxygen of AMI may be the source of free radicals produced from AMI; they therefore synthesized by complete reduction of carbonyl oxygen a synthetic AMI derivate (AMI-R). In contrast to AMI, AMI-R did not exert an acute toxic effect on perfused rabbit lung (pulmonary edema), and was not cytotoxic to human lymphocytes, but retained sodium-channel blocking activity. If this derivate with much less toxicity also possesses the excellent antiarrhythmic effect of AMI, it may have a clinical value [2]. On the in vitro perfused and ventilated rabbit lung model of Kennedy and coworkers, the application of antioxidants prevented pulmonary edema, caused by an increase in permeability, induced by AMI. AMI treatment increased significantly the chemiliminescence of lung tissue in the presence of luminol and also its glutathione disulfide content. On the basis of their results they considered oxidant injury to have a primary role in the pulmonary toxicity of AMI [21].

With these scanty data in mind, our aim was to perform systematic in vitro and in vivo animal experiments and human studies to prove our free radical hypothesis of AMI side effects. Here we deal only with our in vitro studies and in vivo animal experiments.

The concept of the role of free radical reactions in the pathogenesis of AMI side effects is based partly on the chemical structure of the drug and partly on the similarity of side effects of other known free radical generating compounds (nitrofurantoin, ethanol, isoniazide). Structurally AMI is an iodinated benzfurane derivative. The benzfurane compounds are capable of redox reactions and are well-known scavengers of singlet oxygen. The accumulation of lipofuscin – a lipid peroxidation product – in the skin of patients with dermal side effects also stresses the importance of free radical reactions in the drug's side effects. The photosensitizing effect of numerous free radical generating compounds (e.g., phenothiazines) has been proved to be caused by the phototoxic reaction which is mediated by free radicals [13].

Materials and Methods

We used 60 male Wistar rats of 150–200 g body weight for our in vivo and in vitro experiments. We divided them into six groups (control, those treated with amiodarone, amiodarone + MTDQ-DA, amiodarone + silibinin, silibinin, MTDQ-DA), each group containing ten animals. Rats were fed 500 mg/kg body wt. per day AMI for 30 days, added to the standard Lati food by pulverizing amiodarone tablets (Cordarone, Novo Mestro, Krka). We gave antioxidants as a 3-day pretreatment, and then administered antioxidants in parallel to AMI. Both the dihydroquinoline-type MTDQ-DA (6,6'-methylene-bis 2,2 dimethyl-4-methane sulfonic acid: sodium 1,2-dihydroquinoline) [6] and the flavonoid-type silibinin (Madaus, Cologne) [13] antioxidants were administered in a 30–50 mg/kg body wt. per day dose, dissolved in the drinking water of the animals. We checked the daily food and water consumption of the animals, and modified the dosage of drugs according to these results. By the end of the experiments the body weight of animals treated with AMI alone or combined with antioxidants was lower than those not treated with AMI, which indicated a general toxic effect of AMI. We used AMI in a rather high dose in order to safely produce side effects during the relatively short time of the experiments, because our aim was to study the pathogenesis of side effects. Animals were killed by decapitation, their blood collected, and their liver removed. We repeated animal experiments three times, and in each experiment we determined the investigated parameters twice, with 2-2 parallel samples.

Biochemical Studies

The microsomes were prepared by ultracentrifugation. The enzymatically (NADPH and Fe^{3+}) induced lipid peroxidation and the activity of NADPH cytochrome P-450 reductase were studied according to Jordan and Schenkman [20], and the malondialdehyde content (more exact term: thiobarbituric acid reactive substances) was determined by the method of Ottolenghi [30].

The serum malondialdehyde content was also measured by Ottolenghi's method [30], and the liver malondialdehyde content was measured according to Satoh [35]. The protein content of the samples was determined by the method of Lowry et al. [24] using bovine serum albumin as a standard. We measured the (NADPH) cytochrome P-450 reductase (cytochrome c reductase) and NADH cytochrome b_5 reductase (ferricyanide reductase) activities by the method of Jansson and Schenkman [19]. Aminopyrine N-demethylase activity was determined according to Nasch [29]. Serum AST was measured by the kinetically optimized standard method. The AST (28010600) examination was performed by the Baker test with Centrifichem 600 automatic chemical equipment, diluted using a P-1000 pipette.

Chemiluminescence Studies

Light emission was measured using a chemiluminescence method in a CLD-1 Medicor Luminometer with an MMT microprocessor (Medilab, Hungary). During in vitro chemiluminescence measurements we investigated the water suspension of the amiodarone tablet (Cordarone, Novo Mesto, Krka) and amiodarone. A preparation of HCl (Sanofi Pharma, Paris) and an amiodarone injection (Cordarone, Novo Mesto, Krka) were given in the presence of luminol with or without dihydroquinoline-type antioxidants: CH 402 (sodium 2,2-dimethly 1-2,dihydroquinoline-4-yl-methane sulfonate) (USA Patent Nr. 4363910) and MTDQ-DA (6,6'-methylene-bis-2,2-dimethyl-4-methane-sulfonate: sodium 1,2-dihydroquinoline) at changing concentrations of amiodarone and antioxidants. We used divided cuvettes, in the upper part of which we added the material to be examined in a volume of 0.2 ml, the lower part containing the reagent solution. The reagent solution consisted of 0.7 mM luminol, 3.8 μM hemin, and 11.8 mM Na_2CO_3 adjusted to pH 10–11 and bubbled with N_2 gas according to Bögl and Heide [7]. We added the dihydroquinoline-type CH 402 and MTDA-DA antioxidants to the reaction mixture in variable concentrations. The total volume of reagent solution was 1.15 or 1.2 ml. In one experimental series we used the same total volume. After automatic mixing of reagent containing luminol and material to be examined, we measured the intensity of light emission. The chemiluminescence intensity was expressed in millivolts; in this case further data were shown on figures or we used the luminometer in the function mode, which measured the integral value of the light emission's intensity, and in this case we expressed chemiluminescence intensity in millivolts. We performed an average of five measurements from each sample.

For chemiluminescence studies from rat liver homogenate we made a liver homogenate of 3 g/100 ml concentration from the liver of Wistar rats of 150–200 g body weight. The reagent containing luminol described in the previous paragraph was in the lower part, and 50 µl $10^{-3} M$ epinephrine and 20 µl fresh liver homogenate were in the upper part of the cuvette. The protein content of homogenate was referred to unit protein content. We made this by so diluting the suspension of 3 g/100 ml concentration that the protein concentration calculated for total volume would be the same.

Lysosomal Membrane Permeability Studies

The lysosomal enzyme release measurements were performed from the liver of exsanguinated rats. For preparation of liver cell fractions small (1 g weight) pieces of liver tissue were dissolved in an isotonic 0.25 mol saccharose solution, then homogenized with a Teflon-glass potter in an external ice bath, then centrifuged at (4°C at $800 g$ for 15 min. The sediment (M_1 fraction), which contained cell nuclei, cell debris, and cell membrane fragments, was thrown away. The supernatant was further centrifuged at $16 000 g$ for 3 min under similar conditions (at 4°C). The supernatant formed contained the extralysosomal (M_2) fraction and the sediment corresponded to the M_3 fraction rich in lysosomes. This latter was washed twice with isotonic saccharose solution and, after centrifugation with the method described above, we obtained the ML fraction – the sediment of centrifugation at $16 000 g$, which contained the intact liver lysosomes. The aliquot part of the ML fraction was homogenized in the presence of 0.2% Triton X-100 and centrifuged in the above-mentioned way. The supernatant finally obtained, the M_3 fraction, contained the lysosomal enzymes.

For study of lysosomal enzyme release we used 1.0 ml ML fraction dissolved in 0.25 mol saccharose solution, at 37°C, pH = 7.2. After 1 h, incubation samples were immediately cooled to 0°C, and centrifuged at $16 000 g$ for 30 min at 4°C. The released enzyme activity was measured in the supernatant. During study of enzyme release, for each sample we performed an incubation in parallel of 1.0 ml ML fraction in the presence of 0.2% Triton X-100 followed by centrifugation. In the supernatant produced, we measured the total released enzyme activity. In vitro lysosomal membrane permeability was determined by the following formula:

In vitro permeability:

$$Q\ (\%) = \frac{\text{released activity}}{\text{total released activity}} \times 100$$

In vivo lysosomal membrane permeability was assessed by measuring the enzyme activity in M_2 and M_3 fractions, because it was supposed that the lysosomal enzyme activity of the extralysosomal (M_2) fraction was determined in the permeability of the lysosomal membrane and was in equilibrium state in the cells. We neglected lysosomal enzymes reaching the M_2 fraction due to the homogenization process, because every sample was prepared in the same way. In vivo permeability can be expressed as a quotient of extralysosomal in vivo release and the total lysosomal enzyme activity in M_2 and M_3 fractions [41].

In vivo permeability:

$$P\ (\%) = \frac{\text{enzyme activity in } M_2 \text{ fraction}}{\text{enzyme activity of } (M_2 + M_3) \text{ fractions}} \times 100$$

The β-glucuronidase activity was measured according to Barrett [1] at pH = 4.5 and at 37°C; activity was given as that of the unit protein. The protein content of samples was determined according to Lowry [24].

Pulse Radiolysis and Cobalt-60 γ-Radiolysis

Pulse Radiolysis

The computer-controlled pulse radiolysis facility and the fast optical detection system of the Institute of Isotopes has recently been described elsewhere [16]. Briefly, a Tesla-made Linac Model LPR-4 linear accelerator produced 80- to 2600-ns pulses of electrons with a mean energy of 4 MeV. The dose/pulse varied between 10 and 60 Gy, as measured by conventional thiocyanate dosimetry. The formation and decay of transient species was followed by a kinetic spectrophotometer (Applied Photophysics Ltd.) using a xenon arc analyzing light source. Analog/digital conversion, handling and storage of data, and operation of the accelerator were performed by an Iwatsu TS 8123 storage oscilloscope and a Cromemco CS-2H computer. For pulse radiolysis study we used amiodarone substance, which was a generous gift from Sanofi (Paris).

Cobalt-60 γ-Radiolysis

Stationary radiolysis was carried out using the 1600-TBq nominal activity source of the Institute. The dose rate was usually 1100 Gy/h as measured by Fricke dosimetry anc corrected for electron density.

Sample Preparation and Analysis

The chemicals used throughout the study were of analytical grade. The solutions of amiodarone HCl were prepared from doubly distilled water and saturated with appropriate gases (N_2 or N_2O) by gentle bubbling for 30 min before the experiment. The silica optical cells were of 1 cm optical path length and of the flow through type, whenever an effect of the dose was noticed.

Difference spectra of irradiated samples were taken using a Varian DMS-80 spectrophotometer.

The concentration of iodide ion was determined by potentiometry using an iodide ion selective electrode and a calomel reference electrode.

The products of radiolysis were separated by thin-layer chromatography using silica gel plates (Merck Kieselgel 60 F_{254}, 0.2 mm) and a benzene: methonol (3:1) solvent. The irradiated samples were subjected to high-pressure liquid chromatography applying a PIC method, which we developed specifically for the separation of highly polar amino compounds. Analyses were performed on a BST Nevisorb 10-μm C18(250 $\times$ 4 mm) column and a methanol:water (7:3) solvent containing 5 mmol dibutyl amine and phosphoric acid to attain pH = 2.6. The spectrophotometric detector was operated at a wavelength in the region of 200–400 nm.

As standard samples of aminodarone metabolites were not available we prepared partially and fully deiodinated compounds by hydrogenating amiodarone dissolved in methanol using a 10% Pd alumina catalyst.

Morphological Studies

Electron Microscopy

For electron microscopy the small pieces of the rat lung material were fixed with a fixative containing 4% paraformaldehyde and 2.5% glutaraldehyde in phosphate buffer (pH 7.2). After washing, the materials were postfixed in 1% osmic acid for 2 h and embedded in Epon. Ultrathin sections were stained with uranyl acetate and lead citrate. Electron micrographs were taken with a Tesla BS500 electron microscope. For the statistical analysis the number of lysosomes and perisomes were counted within 250–300 μm^2 tissue area from each group and average values for 100 μm^2 tissue area were calculated.

Light Microscopy

We prepared sections by the traditional technique from the liver and lung tissue of rats and stained them with hematoxylin-eosin.

Materials

We purchased glucose-6-phosphate dehydrogenase from Calbiochem (Luzern), CH 402 substance from Chinoin (Budapest), malondialdehyde tetraacetate from Fluka (Buchs), silibinin injection from Madaus Co. (Cologne), epinephrine, osmic acid, uranyl acetate, and lead citrate from Merck (Darmstadt), MIDQ-DA substance from Material Co. (Budapest), amiodarone HCl substance from Sanofi Pharma (Paris), dextran 200, 4-nitrophenyl-β-glucuronide from Serva (Heidelberg), and all other reagents from Reanal (Budapest).

Statistical Analysis

The following statistical analyses were performed: analysis of variance and unpaired or paired Student's t test. When analysis of variance gave a significant difference, we compared the different groups with a modified Student's t test. In some cases we used the confidence limits of the median.

Results and Discussion

Chemiluminescence Measurements

Chemiluminescence-Producing Effect of Amiodarone and the Effect of Antioxidants on Amiodarone-Induced Chemiluminescence

First we investigated the water suspension of Cordarone tablets and Cordarone (Amiodaron) injections in the presence of air. The suspension and solution of 5 and 10 mg/ml concentrations of Cordarone tablets and injections produced a dose-dependent chemiluminescence signal (Fig. 1). The free radical scavenger activity of dihydroquinoline-type antioxidants (CH 402 and MTDQ-DA) used was proved in previous experiments [5, 6]. Adding the above-mentioned dihydroquinoline-type antioxidants to the reaction mixture resulted in a dose-dependent inhibition of AMI-induced chemiluminescence (Fig. 2). Our initial measurements shown in Figs. 1 and 2 indicate only the light intensity changes; later on we were able to use the luminometer in its other function mode to measure the integral value of the intensity light emission. This latter function mode indicated more reliably and quantitatively more exactly the quantity of free radicals produced in the system. When Cordarone injections were studied in this function mode a dose-dependent chemiluminescence signal was observed (see Table 2). In order to exclude the primary role of injection solvent and vehiculum in the tablet in the induction of the chemiluminescence signal, we also studied amiodarone HCl substance, the water suspension of which also produced a dose-dependent chemiluminescence signal (Table 3). When

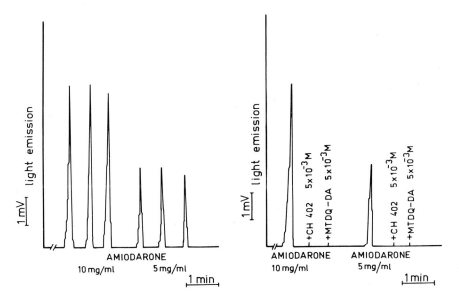

Fig. 1. Chemiluminescence induced by the water solution of amiodarone and the protective effect of added dihydroquinoline-type antioxidants CH 402 and MTDQ-DA. For explanation see text

Table 2. Examination of the free radical generating effect of Cordarone injections (Novo Mesto, Krka) by chemiluminescence measurement in the presence of air

Amiodarone concentration (M)	Light emission (mV)
3.2×10^{-3}	9332.66 ± 192.82 ($n = 6$)
2.55×10^{-3}	7227.25 ± 333.86 ($n = 3$)
1.92×10^{-3}	4720.50 ± 245.23 ($n = 4$)
1.775×10^{-3}	3949.42 ± 1433.67 ($n = 4$)
8×10^{-4}	2617.25 ± 240.43 ($n = 4$)

Measurement conditions: mixing velocity, 3; sensitivity, 1; temperature, 25°C; measurement time, 60 s; total volume, 1150 μl. The average of the measured values was given with a 95% confidence limit: h_1; $h_2 = \bar{x} \pm t_{p5\%} \times s\bar{x}$.

a dihydroquinoline-type antioxidant (MTDQ-DA) was added, a dose-dependent inhibition occurred (Table 4).

The chemiluminescence in our system was caused by excited luminol derivates. Luminol (5-amino-2,3-dihydrophthalazine-1,4-dione) is oxidized to an excited aminophthalate anion by different reactive oxygen intermediates (e.g., $O_2^{\cdot-}$; hydrogen peroxide, H_2O_2; hydroxyl radical, $\cdot OH$, $^1O_2$), which during its return to the ground state emits light [9].

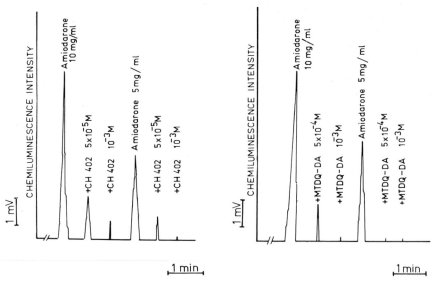

Fig. 2. The dose-dependent inhibition of the chemiluminescence signal induced by the water solution of amiodarone with dihydroquinoline-type antioxidants CH 402 and MTDQ-DA. For explanation see text

Table 3. Study of free radical generating property of amiodarone HCl substance by chemiluminescence measurement in the presence of air

Concentration of amiodarone HCl substance (M)	Light emission (mV)
3.04×10^{-4}	$10\,126.25 \pm 1966.64\ (n = 4)$
2.28×10^{-4}	$7\,569.50 \pm 2561.02\ (n = 3)$
1.52×10^{-4}	$2\,940.16 \pm 2317.22\ (n = 3)$
7.51×10^{-5}	$1\,447.50 \pm 1093.86\ (n = 3)$

Measurement conditions: see Table 2. The average of measured values was given with a 95% confidence limit: h_1; $h_2 = \bar{x} \pm t_{p5\%} \times s\bar{x}$.

Table 4. Study of free radical generating property of amiodarone HCl substance by chemiluminescence measurement in the presence of MTDQ-DA and air

Concentration (M)	Light emission (mV)	
3.04×10^{-4}	Amiodarone HCl substance	$10\,126.25 \pm 1966.64\ (n = 4)$
8.7×10^{-5}	MTDQ-DA	$339.93 \pm 45.28\ (n = 3)$
8.7×10^{-6}	MTDQ-DA	$1\,647.50 \pm 273.26\ (n = 3)$
8.7×10^{-7}	MTDQ-DA	$5\,065.50 \pm 2675.45\ (n = 3)$
4.35×10^{-7}	MTDQ-DA	$8\,307.50 \pm 2326.63\ (n = 3)$
2.6×10^{-7}	MTDQ-DA	$9\,144.15 \pm 1612.31\ (n = 3)$
8.7×10^{-8}	MTDQ-DA	$11\,980.46 \pm 2105.25\ (n = 3)$

Measurement conditions: see Table 2. The average of measured values was given with a 95% confidence limit: h_1; $h_2 = \bar{x} \pm t_{p5\%} \times s\bar{x}$.

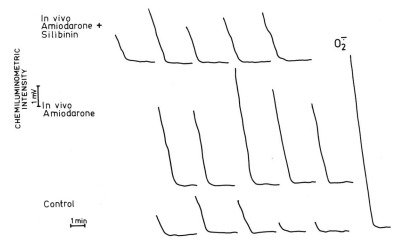

Fig. 3. Effect of in vivo silibinin and amiodarone treatment on the rat liver homogenate O_2^- scavenger function. For explanation see text

The mechanism of chemiluminescence in our system is unclear. AMI may react with O_2 in the presence of air and light, and as a consequence reactive oxygen intermediates (ROIs) or directly $^1O_2$ are generated, which produce chemiluminescence by mediation of luminol, and $^1O_2$ directly as well. However, this is only a possible explanation; further studies are needed to clarify this question.

In summary we proved by in vitro chemiluminescence studies that AMI itself (not just the vehiculum or solvent) generated free radicals in the presence of air and light. As every secondary ROI is able to produce chemiluminescence in the presence of luminol, our studies did not elucidate the mechanism of free radical generation, and the nature of free radicals formed.

Chemiluminescence Study on Liver Homogenate of In Vivo Treated Rats

This study is only a preliminary one, because the method needs further validation. We do not know to what extent the various components of liver homogenate are responsible for the generation of the chemiluminescence signal, which is why it is not certain that light intensity referred to unit protein content or whether the method used in the present experiments is indeed optimal. At that time we could use the luminometer only in the function mode to indicate the amplitude of light intensity. The aim of our study was to determine how the $O_2^{\cdot-}$ scavenger capacity of liver homogenate of animals treated in vivo with AMI or AMI + silibinin changes compared to control $\cdot O_2^{\cdot-}$ was generated by the help of the epinephrine-adrenochrome autooxidative system (Fig. 3). On the right side of Fig. 3, the chemiluminescence signal caused by $O_2^{\cdot-}$ generated during epinephrine-adrenochrome conversion is shown. When the liver homogenate of the control group was added, the intensity of the chemiluminescence signal decreased to a great extent, indicating a significant $O_2^{\cdot-}$ scavenger capacity of control liver homogenate. The addition of liver homogenate of AMI-treated animals decreased the intensity of the chemiluminescence signal significantly less, which means that in vivo AMI treatment essentially decreased $O_2^{\cdot-}$ scavenger capacity of liver homogenate compared to that of the control. The liver homogenate of AMI + silibinin-treated animals had a greater inhibitory effect on the chemiluminescence signal than that of the AMI group, but a smaller inhibitory effect than that of the control. Therefore the simultaneous antioxidant (silibinin) treatment with AMI partially prevented the liver homogenate $O_2^{\cdot-}$ scavenger capacity decreasing the effect of AMI, so that it approached control values, although it did not reach them. The results of this preliminary study also indirectly prove the free radical generating effect of AMI, which can be partially prevented by simultaneous treatment with silibinin.

Biochemical Studies

After obtaining several lines of evidence for free radical generation from in vitro and in vivo chemiluminescence studies, we performed biochemical investigations to examine lipid peroxidation (LPO) induced by free radical reactions, in order to prove indirectly our hypothesis by further independent methods. It is known that chemiluminescence intensity and the accumulation of malondialdehyde (MDA) – an end product of LPO – are proportional to each other, because they are indicators of the same fundamental process, although in different stages. According to Sugioka and Nakano [38], the measured light intensity in the iron-induced LPO of mitochondrial and microsomal membranes was proportional to the square of the accumulated lipid hydroperoxide concentration.

Serum and Liver Tissue Malondialdehyde Content

The serum and liver tissue MDA content of in vivo AMI treated animals was significantly elevated compared to controls. The increase in serum malondialdehyde content was not prevented by antioxidants in the AMI + MTDQ − DA and AMI + silibinin groups; on the other hand the liver tissue MDA content decreased to a value only slightly greater than that of the control in the AMI + MTDQ − DA group, and even below the control value in the AMI + silibinin treated group (Fig. 4). We conclude that our results show increased LPO in sera and liver tissue of AMI-treated animals, but we cannot explain the discrepancy in the effects of antioxidants on serum and liver tissue MDA contents. In spite of the well-known serious limitations of the thiobarbituric acid reaction, we applied this method,

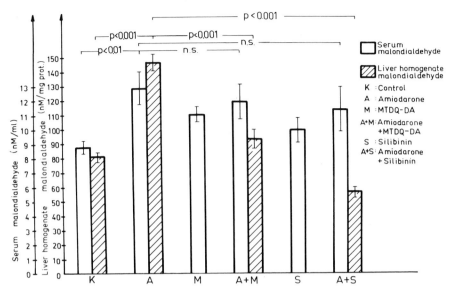

Fig. 4. Serum and liver homogenate malondialdehyde content in rats treated with amiodarone and/or antioxidants. For explanation see text

because we thought that, under controlled circumstances and evaluated together with other independent methods of free radical research, it was a suitable method to investigate free radical reactions, with the restriction that it measured not only serum MDA content but also thiobarbituric acid reactive substances in serum [15].

Induced LPO Measurements

The fundamental conditions for induced LPO to occur are the presence of NADPH, which is necessary for the function of NADPH cytochrome P-450 reductase, and the presence of some iron complex. NADH cytochrome b_5 also plays a role in the electron transport of microsomal LPO. We measured LPO induced by NADPH and Fe^{3+} in vitro, in liver microsomal fraction of in vivo treated animals, by thiobarbituric acid reaction. The enzymatic, induced LPO increased significantly in the AMI-treated group compared to control (Fig. 5). Our previous studies verified that the silibinin and MTDQ-DA antioxidants used in the present experiment were able to inhibit NADPH- and iron-induced LPO of rat liver microsomal fraction [4]. The silibinin and MTDQ-DA antioxidants used were not protective against AMI and caused an increase in the induced, enzymatic LPO.

There is an apparent contradiction between the protective effect of antioxidants against AMI-induced increase of MDA content in liver tissue and the ineffectiveness of antioxidants on induced, enzmyatic liver microsomal LPO caused by treatment with AMI. The explanation of this

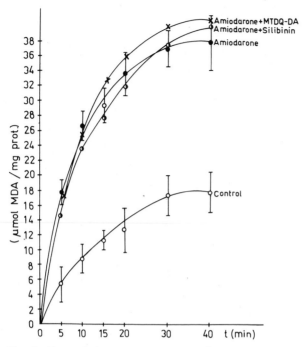

Fig. 5. Comparison of NADPH- and Fe^{3+}-induced microsomal lipid peroxidation in rat microsomes. The figure does not contain the SDs of amiodarone + MTDQ-DA and amiodarone + silibinin groups, becaue it is evident from the figure that there is no significant difference between these two groups, and between these groups and that of amiodarone; therefore in order to avoid confusion caused by many lines in close neighborhood to each other, we have omitted them

apparent contradiction is as follows: during the preparation of microsomal fractions the majority of antioxidants that had access to this site during in vivo treatment were lost; however, in liver homogenate, the antioxidants given access during in vivo treatment were fully present. Consequently the concentration of antioxidants is several orders of magnitude less in liver microsomal fractions than in liver homogenate, which explains the protective effect of in vivo administered antioxidants against AMI-induced LPO in liver homogenate, and its absence in liver microsomal fraction.

Study of Some Enzyme Activities
of the Microsomal Mixed Function Oxygenase System

The enzyme activities of the microsomal mixed function oxygenase system can be considered as indirect indicators of the LPO that occurs during drug metabolism. If, during the drug's metabolism in the liver, microsomal mixed function oxygenase system LPO occurs, a reduction of the enzyme activities examined is to be expected. According to the literature, AMI inhibited

Table 5. Investigation of some enzyme activities of the microsomal oxygenase system in the liver microsomal fraction of rats treated with amiodarone and/or antioxidants

	Aminopyrine N-demethylase (nmol HCHO/mg protxmin)	NADPH cytochrome P-450 reductase (nM reduced cytochrome $\times$ c/mg protxmin)	NADH cytochrome b_5 reductase (mM ferrocyanide/mg protxmin)
Control	5.584 ± 0.073	40.24 ± 0.40	3.13 ± 0.23
Amiodarone	5.464 ± 0.227	48.34 ± 1.59	3.57 ± 0.14
Amiodarone + silibinin	5.627 ± 0.066	52.05 ± 2.12	3.82 ± 0.10
Amiodarone + MTDQ-DA	5.226 ± 0.163	50.99 ± 7.43	3.93 ± 0.32

The data shown are triplicate measurements of enzyme activities with two parallels, prepared from the liver homogenate of 5-5 animals in each group.

Table 6. Serum AST values of in vivo treated rats

Group	$\bar{x}$ ± SEM	n
Control	75.6 ± 7.3	10
Amiodarone	120.1 ± 15.4	10
MTDQ-DA	123.3 ± 8.2	8
Amiodarone + MTDQ-DA	101.0 ± 9.2	10
Silibinin	103.1 ± 3.9	9
Amiodarone + silibinin	82.1 ± 7.9	9

The line (amiodarone) and column (treatment) effect was not significant according to the two-way analysis of variance.

Table 7. In vivo β-glucuronidase permeability of rat liver lysosomal membrane

Group	In vivo permeability $\% = \dfrac{M_2}{M_2 + M_3} \times 100$
Control ($n = 8$)	13.61
Amiodarone ($n = 8$)	31.87
Amiodaron + MTDQ-DA ($n = 8$)	22.46
Amiodarone + silibinin ($n = 8$)	17.81

the aminopyrine-*N*-demethylase and aniline-hydroxylase activities, which were not the consequences of a direct inhibitory action of AMI, but a consequence of diminished cytochrome P-450 and cytochrome b_5 content [3]. We examined the aminopyrine-*N*-demethylase, NADPH cytochrome P-450 reductase, and NADH cytochrome b_5 reductase activities in the liver microsomal fraction of in vivo treated rats. Neither AMI treatment nor AMI and simultaneous antioxidant (silibinin, MTDQ-DA) treatment influenced the above-mentioned enzyme activities in comparison with the control. It may be that, in spite of the rather high dose, the AMI did not reduce the enzyme activities examined during these relatively short term experiments (Table 5). Our previous investigations verified that the antioxidants used did not influence in themselves the examined enzyme activities [4], which is why in this experiment we did not use groups treated with only silibinin or MTDQ-DA.

Serum Aspartate Aminotransferase Determination

There was no measurable difference between the serum aspartate amino-transferase (AST) values of different in vivo treated groups according to the analysis of variance (Table 6). AMI treatment causes relatively frequently mild elevation of serum aminotransferase values in patients (15%–20% of patients treated), but it usually manifests itself after 2–4 months of treat-ment. More serious liver injury (hepatitis, cirrhosis) was rarely observed; it was not dose dependent [42]. In our animal experiments the treatment time was probably too short to induce serum AST level elevation or more serious injury in liver tissue with a very large antioxidant supply.

Lysosomal Membrane Permeability Studies

The lysosomal damaging effect of free radical reactions is well established [39, 8, 12]. The increased release of lysosomal enzymes as a consequence of increase in lysosomal membrane permeability is an indicator of lysosomal damage by free radical reactions; we performed the above-mentioned exam-inations in order to indirectly reveal AMI-induced free radical generation.

The in vitro and in vivo β-glucuronidase permeability of in vivo treated rat liver lysosomes was significantly increased in the AMI treated group compared to control. The increase of in vivo lysosomal permeability was significantly diminished by simultaneous MTDQ-DA or silibinin treat-ment with AMI, the application of the latter antioxidant decreasing in vivo lysosomal permeability almost to control values (Fig. 6, Table 7).

Figure 6 shows that the β-glucuronidase activity of the extralysosomal fraction (M_2) increased to a large extent on AMI treatment compared to the control, and the preventive effect of simultaneous MTDQ-DA and silibinin treatment was also demonstrated. According to the increased β-glucuronidase release in the AMI-treated group compared to control, in the lysosomal

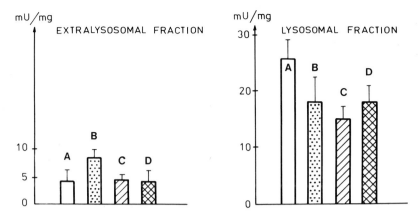

Fig. 6. The β-glucuronidase activity of rat liver lysosomal (M₃) and extralysosomal (M₂) fraction. [A], control; [B], amiodarone; [C], amiodarone + MTDQ-DA; [D], amiodarone + silibinin. For further explanation see text

(M_3) fraction the remaining enzyme activity was lower than that of the control. Table 6 demonstrates in vivo permeability values calculated on the basis of enzyme activity data in M_2 and M_3 fractions.

Neither antioxidants could prevent the large increase of in vitro lysosomal membrane permeability caused by AMI. Similar trends of in vitro lysosomal β-glucuronidase release were observed in isoosmosis and hypoosmosis, although of course in the latter case we obtained greater absolute values (Fig. 7).

Morphological Studies

Light Microscopic Results

Light microscopic sections were performed from rat liver and lung tissue. Sections were designated by code numbers, so the person who evaluated them did not know the group they belonged to. Examination of the liver tissue revealed characteristic alterations. Parenchymal and/or fatty degeneration of the liver was observed in several cases, but these could not be related to any treatment, because they occurred in every group, with approximately the same frequency. This fact was well matched with our serum AST results, because there was no significant difference between the serum AST values of the different groups.

Study of the lung tissue in the groups treated with AMI alone or AMI with one of the antioxidants showed almost uniformly "foamy cells," which corresponded to alveolar macrophages. The "foamy cytoplasm" is the light microscopic equivalent of lysosomal phospholipidosis, which can be seen by electron microscopic examination. The mildest alteration was the slight

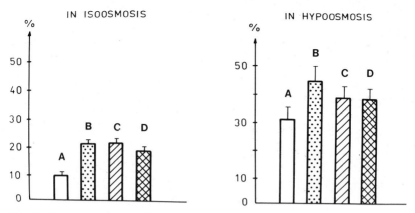

Fig. 7. In vitro β-glucuronidase release from rat liver lysosomes in hypoosmotic and isoosmotic environments. ⒜, control; ⒝, amiodarone; ⒞, amiodarone + MTDQ-DA; ⒟, amiodarone + silibinin. For further explanation see text

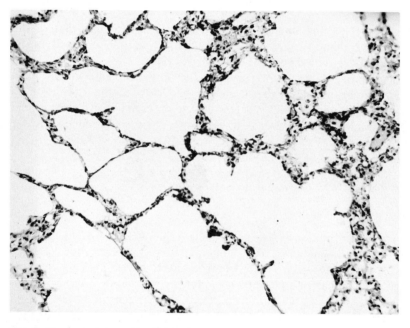

Fig. 8. Rat lung tissue under light microscopy. Accumulation of "foamy cells" with slight thickening of the interalveolar septa

accumulation of "foamy cells" in interalveolar septa. However, in the majority of animals more serious alterations were found: moderate or significant accumulation of "foamy cells" with the thickening of alveolar septa, pneumonitis to differing extents, and graveness (Figs. 8–11).

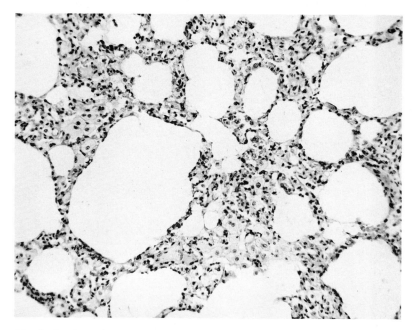

Fig. 9. Rat lung tissue under light microscopy. Significant accumulation of "foamy cells" with a greater degree of thickening of the interalveolar septa

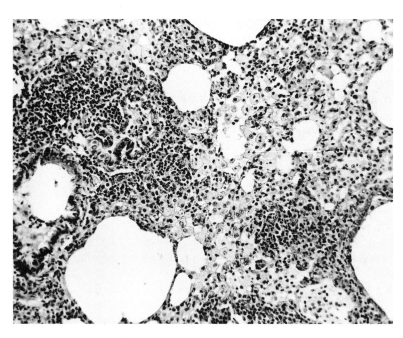

Fig. 10. Rat lung tissue under light microscopy. Significant accumulation of "foamy cells" and conspicuously thickened interalveolar septa, with inflammatory small cell infiltrate (pneumonitis)

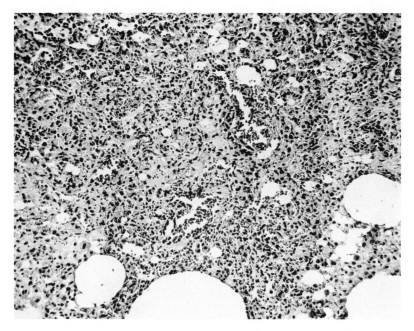

Fig. 11. Rat lung tissue under light microscopy. Largely identical to Fig. 10, but granulation tissue, fibrosis, and signs of consolidation also present

For semiquantitative evaluation of morphological alterations in the lung, we created the concept of "pathological lung" index, which was determined by the quantity of "foamy cells" (designated by $+-++++$ in order of graveness), and the percentage of lung tissue involved in the above-mentioned morphological alterations within the total area of lung tissue in the section, according to the following formula:

$$\text{"Pathological lung" index} = \text{number of} +/-\text{s denoting foamy cells} \times \text{pathological area (\%)}$$

All sections were evaluated at least twice, and our semiquantitative evaluation method proved to be well reproducible. The values of "pathological lung" index are demonstrated in Table 8. It is striking that highest values were found in the three AMI-treated groups. The simultaneous antioxidant treatment could not influence the "pathological lung" index compared to the group treated with AMI alone. The values of the AMI-treated groups were significantly higher than those of the groups not treated with AMI. The "pathological lung" index of groups treated with only an antioxidant (MTDQ-DA or silibinin) was lower than the control, but it was not statistically significant. Our light microscopic results demonstrated above are not new, they just confirm the results of several other authors [10, 23]. The purpose of the light microscopic examinations was to prove that the dosage

Table 8. "Pathological lung" index values determined by light microscopy from rat lung tissue in the different treated groups

Group	"Pathological lung" index = the number of +/−s denoting foamy cells × pathological area (%) "Pathological lung" index = $\bar{x} \pm$ SEM
Control ($n = 10$)	87.5 ± 36.14
MTDQ-DA ($n = 9$)	9.45 ± 4.50
Amiodarone ($n = 10$)	127.0 ± 48.47
Amiodarone + MTDQ-DA ($n = 10$)	142.0 ± 34.77
Silibinin ($n = 8$)	7.51 ± 4.12
Amiodarone + silibinin ($n = 10$)	152.0 ± 42.91

The line (amiodarone) effect is highly significant ($p < 0.001$) according to the two-way analysis of variance, which means that the values of the amiodarone-treated groups are significantly different from those of groups not treated with amiodarone. Although the "pathological lung" index values of MTDQ-DA- and silibinin-treated groups seem to be essentially lower than those of controls, this difference was not statistically significant. The simultaneous antioxidant treatment with amiodarone did not influence the "pathological lung" index compared to the group treated with amiodarone alone.

and time of AMI treatment were sufficient to induce the drug's pulmonary side effects in our experiments, and by that we could ensure that other studies (biochemical, electron microscopic) performed with the aim of elucidating the pathogenesis of the drug's side effects were performed in animals in which the drug was certain to induce some side effects.

Electron Microscopic Results

Electron micrographs were taken from rat liver and lung tissue. Being aware of the light microscopic results we expected the electron microscopic study to show a different picture. The lysosomal phospholipoidosis which can be seen by electron microscopy has been described by many authors [27, 34, 36] in the organs of AMI-treated animals and patients. Our aim was not merely to confirm these results, but to find out whether antioxidant treatment could influence lysosomal phospholipoidosis, and thus we attempted to obtain indirect data on the possible relationship between free radical reactions and lysosomal phospholipoidosis. As far as we know such a study has not yet been carried out.

First we present the results of the lung tissue examination. Figure 12 shows the electron micrograph of an alveolar epithelial cell from the control group. Under the electron microscope the alveolar epithelial cells possess short microvilli on their free surface and form junctional complexes with neighboring alveolar epithelial cells. There are abundant free ribosomes, vesicular and cisternal profiles of the granular endoplasmic reticulum, Golgi complex, and some dense small lysosomes 0.3–1 μm in diameter. They have an internal structure consisting of thin parallel or concentric lamellae and

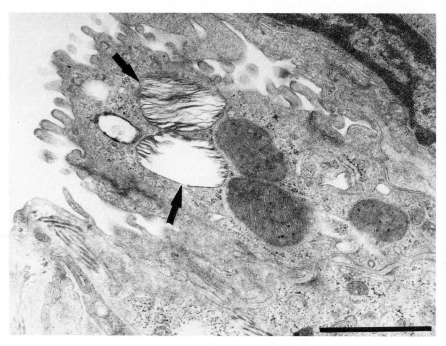

Fig. 12. Electron micrograph of an alveolar epithelial cell in the control animal. *Arrows* show the lysosomes in the cytoplasm. Scale bar = 1 μm, ×42 000

are limited by a membrane. They are distributed throughout the cytoplasm of the alveolar epithelial cells.

After treatment with AMI electron microscopy showed intralysosomal myelin figures, suggestive of phospholipoidosis. These myelin figures were associated with intralysosomal electron deposits. The myelin-containing lysosomes were round, ovoid, or irregular in shape, they were enlarged, and their diameter varied from 0.5 to 2 μm. Their number was increased compared to the control (Figs. 13, 14).

The number of lysosomes was decreased after silibinin treatment, although a large number of lysosomes with intralysosomal myelin figures and electron-dense deposits such as those observed in AMI-treated animals were found (Fig. 15).

In order to support our qualitative observations, we also used a semiquantitative approach: after surveying a 300-μm² tissue area in each group, we calculated the number of lysosomes and perisomes for a 100-μm² tissue area in each, as shown in Table 9. The semiquantitative results supported our qualitative impressions: treatment with AMI significantly increased the number of lysosomes, and the simultaneous silibinin treatment diminished to a large extent the increased lysosome number caused by AMI treatment.

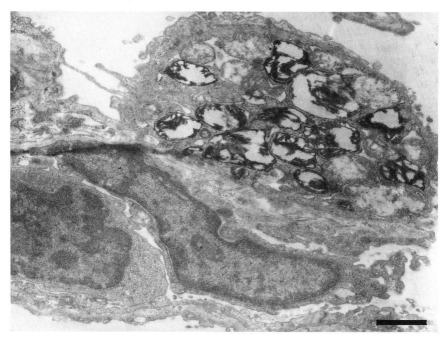

Fig. 13. After amiodarone treatment the number and size of the lysosomes increased in the alveolar epithelial cell. Scale bar = 1 μm, ×18 600

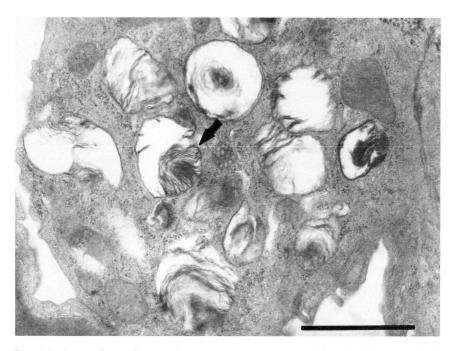

Fig. 14. *Arrow* shows the membranous structures arranged in whorled arrays in the lysosomes. Scale bar = 1 μm, ×42 000

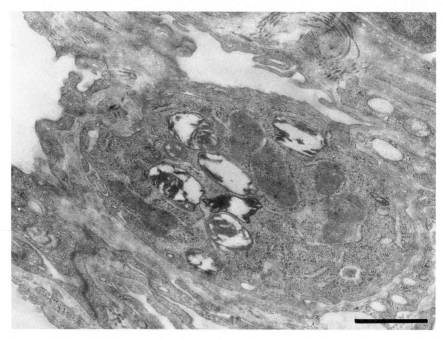

Fig. 15. After silibilin treatment the number of lysosomes is decreased compared with amiodarone treatment. Scale bar = 1 μm, ×24 000

Table 9. Number of lysosomes calculated for 100 μm² tissue area

Group	$\bar{x} \pm$ SEM		
Control ($n = 8$)	7.3 ± 1.4	$p < 0.001$	
Amiodarone ($n = 8$)	26.2 ± 2.8		
Amiodarone + silibinin ($n = 8$)	13.7 ± 2.5	$p < 0.01$	$p < 0.05$
Amiodarone + MTDQ-DA ($n = 8$)	23.4 ± 3.8	ns	

The simultaneous MTDQ-DA treatment with AMI was ineffective in this respect.

Our results suggest the very important indirect conclusion that the simultaneous administration of silibinin with AMI significantly decreased the accumulation of lysosomal phospholipids – since the decrease in lysosome number means that the absolute quantity of lysosomal phospholipoidosis was also diminished – but it could not entirely prevent the lysosomal phospholipoidosis. This supports the concept that free radical reactions may be involved in the development of lysosomal phospholipoidosis.

We also found the above-mentioned characteristic electron microscopic alterations in the liver tissue, but to a lesser degree and less reproductively

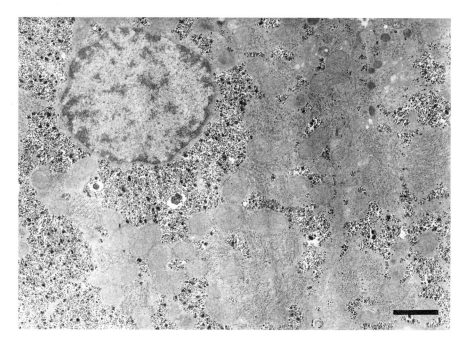

Fig. 16. Electron micrograph of a normal liver cell. Large quantity of glycogen granules in the cytoplasm, rough endoplasmic reticulum, and mitochondria are seen. Scale bar = 1 μm, ×18 000

than in the lung; therefore we could not carry out a suitable study of the effect of antioxidants (Figs. 16–18).

Pulse Radiolysis and Cobalt-60 γ-Radiolysis Experiments

The reactivity of AMI toward reducing agents was examined using stationary cobalt-60 γ- and electron-pulse -radiolysis. On irradiating a dilute aqueous solution of an alcohol containing nitrous oxide, reducing free radicals (solvated electron = e_{aq}^- or α-hydroxy radicals = $\cdot OH$) can be produced as shown by Eqs. 1–3:

$$H_2O \rightsquigarrow e_{aq}^-, \cdot OH \tag{1}$$

$$e_{aq}^- + N_2O \rightarrow N_2 + \cdot OH + OH^- \tag{2}$$

$$\cdot OH + R_1R_2CHOH \rightarrow H_2O + R_1R_2 \cdot COH \tag{3}$$

In the presence of AMI ($1 \times 10^{-3} M$) at neutral pH, formation of iodide ion was observed by potentiometry using an iodide ion selective electrode (Fig. 19).

Using various alcoholic solutions the yield of iodide ion increased in the order:

tert-Butanol<methanol<ethanol<isopropanol

Fig. 17. After amiodarone treatment the glycogen granules completely disappeared, in the cytoplasm there was a conspicuous accumulation of smooth endoplasmic reticulum, and lysosomes with myelin figures appeared (*arrow*). Scale bar = 1 μm, ×44 000

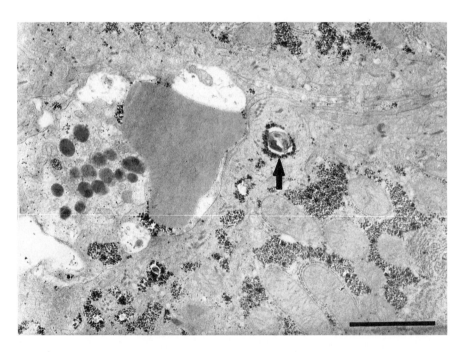

Fig. 18. After the simultaneous silibinin treatment with amiodarone the glycogen granules appeared again in the cell; however, the number of lysosomes decreased. The lysosome in the cytoplasm is shown by the *arrow*. Scale bar = 1 μm, ×36 000

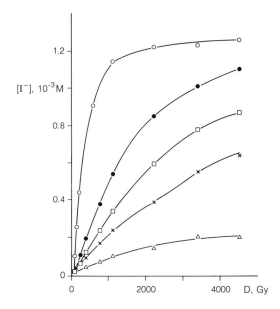

Fig. 19. Concentration of I in aqueous alcohol (30%, v/v) solutions containing 1 mM AMI, γ-irradiated with various doses, D. ○, iso-propanol, bubbled with N$_2$O; ◇, ethanol, N$_2$O; □, methanol, N$_2$O; X *tert*-butanol, N$_2$; △, *tert*pbutanol, N$_2$O

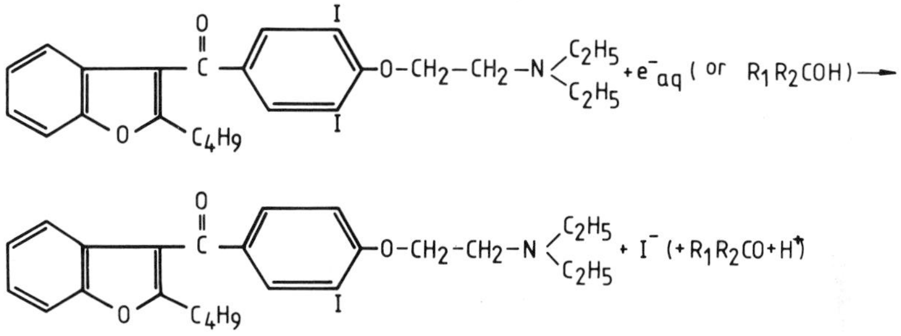

Fig. 20. Formation of aryl radical from amiodarone in reducing environment. e_{aq}^{-a}, solvated electron; R$_1$R$_2$COH, α-hydroxy radical situated on an organic molecule; R$_1$R$_2$CO, carbonyl group situated on an organic molecule; H$^+$, hydrogen ion. For further explanation see text

The concentration of iodide ion (I$^-$) in isopropanol/water irradiated with doses greater than 1 kGy in the presence of 1 mM AMI was well in excess of 1 mmol, indicating that at higher conversion of AMI full deiodination can take place, by releasing both iodine atoms in succession. Analyses in the early stages of the reaction, at low doses of stationary or electron-pulse irradiation, confirmed that partially deiodinated AMI was the major product in each case. On reduction of AMI, I$^-$ is released, accompanied by a simultaneously formed aryl radical (Fig. 20). The aryl radical is highly reactive, readily abstracting hydrogen atoms from neighboring organic

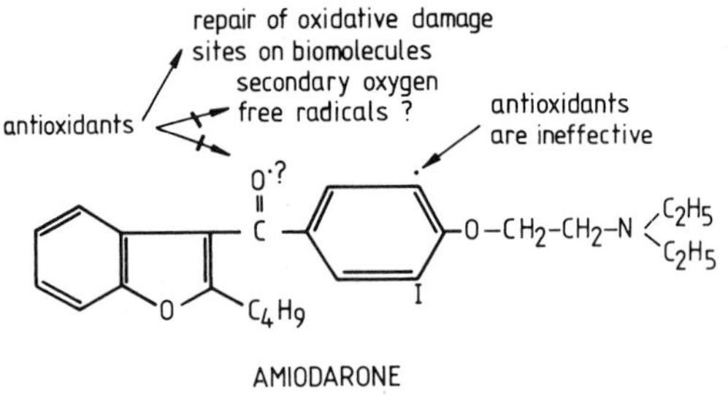

Fig. 21. In a reducing environment a very reactive aryl radical is formed from amiodarone. The formation and damaging effects of the aryl radical cannot be prevented by antioxidants. A less reactive carbonyl radical may be formed also from amiodarone according to Bennett et al. [2]. We found no data to support this suggestion but, if carbonyl radical is formed, being less reactive, it may be scavenged by antioxidants. The other basis of the partial protective of antioxidants effect against amiodarone toxicity may be the repair of oxidatively damaged biomolecules and the scavenging of the probably generated secondary oxygen free radicals. The figure shows carbonyl radical and aryl radical at the same time on an amiodarone molecule, but only one of them is formed by the reduction of one amiodarone molecule. +, inhibition

molecules, thereby intitiating LPO or other free radical reactions. We verified this in vitro in a solution containing isopropanol, which is known to have a labile hydrogen atom in the molecule; aryl radical was capable of carrying out a chain reaction.

Our pulse radiolysis results can probably be extrapolated to in vivo circumstances, and the aryl radical detected in vitro may also emerge in vivo during AMI metabolism in microsomal drug metabolizing enzyme systems. The very reactive aryl radical reacts at once at its site of production with neighboring biomolecules; antioxidants are probably unable to prevent this reaction; they may just be capable of partially repairing the oxidative damage sites produced, or of scavenging secondary ROIs if they are produced (Fig. 21). According to our results AMI reacts rapidly with solvated electrons (e_{aq}^-) (a strong reducing agent), and the reaction is diffusion controlled. With other weaker reducing agents AMI reacts about 3 orders of magnitude slower, which may explain why AMI toxicity usually manifests itself after a relatively longer treatment, because the reactive aryl radical probably forms slowly in vivo (results not shown here, we intend to publish them in a separate publication).

In various alcoholic solutions of AMI a stable product absorbing at 370 nm is formed as a consequence of stationary cobalt-60 γ- or electron-pulse irradiation under reducing conditions. We tried to identify this stable product by thin-layer chromatography and high-pressure liquid chromatography. As standard samples of AMI metabolites necessary for the evaluation of chromatograms were not available, we prepared partially and fully

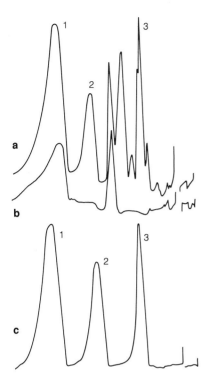

Fig. 22a–c. High-performance liquid chromatogram of alcoholic AMI solution exposed to various reducing conditions. **a** 1 mM AMI in 30% (v/v) isopropanol/water saturated with N_2O and irradiated with a dose of 550 Gy. Detection at 242 nm. **b** Same solution as in **a**, detected at 370 nm. **c** 1 mM AMI in methanol, subjected to catalytic hydrogenation, detected at 243 nm. *Peak 1* is assigned to AMI, *peak 2* to partially, and *peak 3* to fully, deiodinated AMI

deiodinated compounds by hydrogenating AMI dissolved in methanol using a 10% Pd/alumina catalyst. Figure 22, chromatogram b, shows the stable product absorbing at 370 nm, which is also generated beside the main product (the partially deiodinated AMI). The stable product was not successfully identified. It is obvious from Fig. 22 that the stable product is different from the deiodinated products prepared by catalytic hydrogenation of AMI. An undesirable property of the stable product, its decomposition within a couple of days even in the dark, greatly impedes its identification. The stable product might be formed in recombination processes of free radicals generated in the system.

Bennett and coworkers [2] suggested the formation of carbonyl radical on the AMI molecule; we cannot exlude this possibility, but found nothing to support the concept. The generation of secondary ROIs was also supposed from AMI after photolysis, and further studies are needed to clarify whether secondary ROIs are also formed in our system [18, 23, 31].

As we mentioned in respect of the main products, similar results were obtained in photolysis studies by several authors [31], and they suggested the role of the generated aryl radical and secondary ROIs in the pathogenesis of AMI-induced photosensitivity. The significance of our study design is that, as far as we know, the effect of the reducing environment produced by stationary cobalt-60 γ-irradiation or pulse radiolysis on AMI has not yet been investigated. This is important, because during the metabolism of AMI in vivo in the liver microsomal electron transport chain AMI may meet this reductive environment, and in the organism under both physiological and pathological conditions free radicals with a reducing property may occur, which when they react with AMI are able to induce the reactions that took place in our experimental system.

Conclusions

The primary purpose of our study was to prove the essential role of free radical reactions in the pathogenesis of AMI toxicity. We verified indirectly and directly with our systematic in vitro and in vivo chemiluminescence, biochemical, lysosomal membrane permeability, morphological, and physicochemical studies that AMI in vitro and very probably also in vivo generates free radicals, and the oxidative stress induced by AMI was intimately involved in the pathogenesis of side effects. It is important to emphasize that the finding of free radical generation by AMI seems to be well based by many independent methods, which indicated different stages of free radical chain reactions.

A very important finding was that the simultaneous antioxidant (silibinin) treatment was able to decrease significantly the AMI-induced increase of lysosome number, and consequently to decrease the lysosomal phospholipoidosis, considered to be a specific marker of AMI (and other amphiphilic drug) induced organ damage, although it could not prevent lysosomal phospholipoidosis. There may be multiple relationships between free radical reactions and lysosomal phospholipoidosis: the lysosomal damaging effect of free radical reactions is well known, our lysosomal membrane permeability studies indirectly supporting this by demonstrating lysosomal damage caused by AMI-induced free radicals. There are several lines of evidence in the literature that free radical reactions are able to produce morphological alterations similar to lysosomal phospholipoidosis [38, 40]; it was also shown by other authors that lipofuscin – an LPO product – at least partly, was a component of accumulated lysosomal phospholipids [8]; however, further studies are necessary to clarify the possible relationships.

Our physicochemical studies not only supported the view of free radical generation by AMI, but also identified the structure and mechanism of formation of the main free radical product, which may also be produced in vivo. This also explains the discrepancy that in vitro the antioxidants were effective against AMI-induced alterations, but their in vivo protective effect was ambiguous. As we indicated, antioxidants were unable to scavenge even in optimal conditions the very reactive aryl radical; their ability to scavenge secondary ROIs and to repair oxidative damage on biomolecules may account for their partial protective effect.

The more conspicuous damage of lung tissue by AMI, compared to the mild alterations found in liver tissue, needs further investigation; we think that it may be caused by the better antioxidant supply of liver tissue, and the significantly greater accumulation of AMI in lung tissue than liver tissue in rats [22].

The free radical concept uniformly explains the pathogenesis of all AMI side effects, and the only exception seems to be the thyroideal side effects. These are caused partly by the high iodine content of the AMI molecule, and it is also known that AMI probably induces "selective cardial hypothyreosis" by inhibiting the myocardial effect of triodothyronine (T_3) at a cellular level. Due to its similar structure to thyroideal hormones, AMI inhibits the binding of T_3 to its nuclear receptor and the uptake of thyroxine (T_4) into the cell, as well as the peripheral $T_4 \rightarrow T_3$ conversion [42].

Although the simultaneous administration of antioxidants with AMI seems to be justified – provided they do not abolish the drug's main effect – in order to diminish the incidence and degree of AMI toxicity the synthesis of an AMI derivative with the same antiarrhythmic effect, but less free radical generating property, may be a better approach to this problem, because the protective effect of antioxidants could be only partial even under optimal conditions.

References

1. Barrett AJ (1972) Assay method of lysosomal enzymes. In: Lysosomas, a laboratory handbook. Dingle Amsterdam
2. Bennett PB, Kabalka G, Kennedy TP, Woosley RL, Hondeghem LM (1987) An amiodarone derivative with reduced toxicity and Na-channel blocking properties. Circulation 76 Suppl IV:IV-150
3. Berger Y, Harris L (1986) Pharmacokinetics. In: Harris L, Roncucci R (eds) Amiodarone. Pharmacology – pharmacokinetics – toxicology, clinical effect. MEDSI, Paris, pp 46–98
4. Blázovics A, Somogyi A (1988) The role of free radical reactions in experimental hyperlipidemia and athero-sclerosis. Thesis, Budapest.
5. Blázovics A, Somogyi A, Lengyel G, Láng I, Fehér J (1988) Inhibition of lipid peroxidation by dihydroquinoline-type antioxidant (CH 402). Free Radic Res Commsun, 4:409–413
6. Blázovics A, György I, AJN Zsinka A, Biacs P, Földiák G, Fehér J (1989) In vitro scavenger effect of dihydroquinoline type derivates in different free radical generating systems. Free Radic Res Commun 6:217–226
7. Bögl W, Heide L (1985) Chemiluminescence measurements as an identification method for gamma-irradiated foodstuffs. Radiats Fiz Chem 25:173–185
8. Brunk U (1988) The potential intermediate role of lysosomes in oxygen free radical pathology. APMIS 96:3–13
9. Cadenas E, Sies H (1984) Low level chemiluminescence as an indicator of singlet molecular oxygen in biological systems. Methods Enzymol 105:221–231
10. Costa-Jussa FR, Corrin B, Jacobs JM (1984) Amiodarone lung toxicity: a human and experimental study. J Pathol 143:73–79

11. Counihan PJ, Mc Kenna WJ (1989) Low dose amiodarone for the treatment of arrhythmias in hypertrophic cardiomyopathy. J Clin Pharmacol 29:436–438
12. Fehér J, Toncsev H, Fehér E, Kiss Á, Vasadi Á (1981) Lysosomal enzymes in sera and granulocytes of patients with chronic liver diseases. Int J Tissue React III (1):31–37
13. Fehér J, Csomós G, Vereckei A (1987) Free radical reactions in medicine. Springer, Berlin Heidelberg New York
14. Guigui B, Perrot S, Berry JP, Fleury-Feith J, Martin N, Métreau JM, Dhumeaux D, Zafrani ES (1988) Amiodarone-induced hepatic phospholipoidosis: a morphological alteration independent of pseudoalcoholic liver disease. Hepatology 8:1063–1068
15. Gutteridge JMC (1987) Lipid peroxidation: some problems and concepts. In: Oxygen radicals and tissue injury. Proceedings of a Brook Lodge Symposium Augusta, Michigan, USA, April 27–29, 1987, p 9–19
16. György I, Földiák G (1988) Formation and decay of phenoxyl radicals: variation with p and pulse dose. J Radioanal Nucl Chem 122:207
17. Harris L, Michat L (1986) Clinical efficacy-arrhythmias In: Harris L, Roncucci R (eds) Amiodarone. Pharmacology-pharmacokinetics-toxicology-clinical effects. MEDSI, Paris, pp 137–162
18. Hassan T, Kochevar IE, Abdulah D (1984) Amiodarone photoxicity to human erythrocytes and lymphocytes. Photochem Photobiol 40:715–719
19. Jansson J, Schenkman JB (1977) Studies on three microsomal electron transfer enzyme systems (specificity of electron flow pathways). Arch Biochem Biophys 178:89–107
20. Jordan RA, Schenkman JB (1982) Relationship between malondialdehyde production and arachidonate consumption during NADPH-supported microsomal lipid peroxidation. Biochem Pharmacol 31:1393–1400
21. Kennedy TP, Gordon GB, Paky A, Mc Shone A, Adkinson NF Jr, Peters SP, Friday K, Jackman W, Sciuto AM, Gurtner GH (1988) Amiodarone causes acute oxidant lung injury in ventilated and perfused rabbit lungs. J Cardiovasc Pharmacol 12:23–36
22. Latini R, Bizzi A, Cini M, Veneroni E, Marchi S, Riva E (1987) Amiodarone and desethylamiodarone tissue uptake in rats chronically treated with amiodarone is non-linear with the dose. J Pharm Pharmacol 39:426–431
23. Li ASW, Chignell CF (1987) Spectroscopic studies of cutaneous photosensitizing agents – IX. A spin trapping study of the photolysis of amiodarone and desethylamiodarone. Photochem Photobiol 45:191–197
24. Lowry AH, Rosenbrough NJ, Farr AL, Randall RJ (1951) Protein measurement with the Folin-phenol reagents. J Biol Chem 193:265–275
25. Martin WJ, Rosenow EC (1988a) Amiodarone pulmonary toxicity. Recognition and pathogenesis (part 1). Chest 93:1067–1075
26. Martin WJ, Rosenow EC (1988b) Amiodarone pulmonary toxocity. Recognition and pathogenesis (part 2). Chest 93:1242–1248
27. Mason JW (1987) Amiodarone. N Engl J Med 316:455–466
28. Nademanee K, Stevenson W, Weiss J, Singh BN (1988) The role of amiodarone in the survivors of sudden arrhythmic deaths. In: Singh BN (ed) Control of cardiac arrhythmias by lengthening repolarization. Futura Mount Kisco, New York, pp 429–508
29. Nasch T (1953) The colorimetric estimation of formaldehyde by means of the Hantzsch reaction. Biochem J 55:416–421
30. Ottolenghi A (1959) Interaction of ascorbic acid on mitochondrial lipids. Arch Biochem Biophys 79:355–363
31. Paillous N, Verrier M (1988) Photolysis of amiodarone, an antiarrhythmic drug. Photochem Photobiol 47:337–343
32. Rakita L, Sobol SM, Mostow N, Vrobel T (1983) Amiodarone pulmonary toxicity. Am Heart J 106:906–915
33. Ratliff NB, Estes ML, Myles JL, Shirey EK, McMahon JT (1987) Diagnosis of chloroquine cardiomyopathy by endomyocardial biopsy. N Engl J Med 316:191–193

34. Rigas B, Rosenfeld LE, Barwick KW, Enriquez R, Helzberg J, Batsford WP, Josephson ME, Riely CA (1986) Amiodarone hepatotoxicity. Ann Intern Med 104:348–351
35. Satoh K (1978) Serum lipid peroxide in cerebrovascular disorders determined by a new colorimetric method. Clin Chim Acta 90:37–43
36. Simon JB, Manley PN, Brien JF, Armstrong PW Amiodarone hepatotoxicity simulating alcoholic liver disease. N Engl J Med 311:167–172
37. Somani P (1989) Basic and clinical pharmacology of Amiodarone: relationship of antiarrhythmic effects, dose and drug concentrations to intracellular inclusion bodies. J Clin Pharmacol 29:405–412
38. Sugioka K, Nakano M (1976) A possible mechanism of the generation of singlet oxygen in NADPH-dependent microsomal lipid peroxidation. Biochem Biophys Acta 423:213–216
39. Tappel AL (1973) Lipid peroxidation damage to cell components. Fed Proc 32:1870–1874
40. Tigyi A, Zsoldos T, Montskó T (1984) The pathogenesis of experimental silicosis "Free radicals and tissue damage". Scientific sesssion. Pécs, 10–11, January 1984
41. Toncsev H, Frenkl R (1984) Studies on the lysosomal enzyme system of the liver in rats undergoing swimming training. Int J Sports Med 5:152–155
42. Vrobel TR, Miller PE, Mostow ND, Rakita L (1989) A general overview of amiodarone toxicity: its prevention, detection, and management. Prog Cardiovasc Dis 31:393–426

A handy reference book – up to date!

J. Prieto, University of Navarra, Pamplona, Spain;
J. Rodés, University of Barcelona, Spain;
D. A. Shafritz, Albert Einstein College of Medicine,
Bronx, NY (Eds.)

Hepatobiliary Diseases

1992. Approx. 900 pp. 236 figs. 100 tabs.
Hardcover. ISBN 3-540-54326-0

Liver diseases are very prevalent in the general population,
and knowledge of their causes, mechanisms, diagnostic
techniques, and treatment has expanded enormously in
recent years. This has created a real need for a comprehen-
sive account of modern hepatology, and this book meets
that need. It covers the entire spectrum of hepatobiliary
diseases, includes chapters on molecular biology of the
liver, and deals in detail with new areas such as hepatitis C
and E and magnetic resonance
imaging. The authors have been
drawn from all over the world and
are recognized experts in their
subjects. The balanced basic and
clinical approach, combined with a
very direct and lucid style of writ-
ing, ensures that the book will be
of great value to both students and
senior clinicians.

Springer-Verlag
Berlin
Heidelberg
New York
London
Paris
Tokyo
Hong Kong
Barcelona
Budapest